AGRICULTURAL TECHNOLOGIES

AGRICULTURAL TECHNOLOGIES

– Compiled and Edited by –

Dr. Shahid Ahamad
Associate Director Research

Dr. R.K. Samnotra
Director Research

Dr. Pardeep Wali
Ex., Director Research

Directorate of Research

Sher-e-Kashmir University of Agricultural Science and Technology of Jammu

ASTRAL
Publishing Authors Across the Globe

2025

Daya Publishing House®
A Division of
Astral International Pvt. Ltd.
New Delhi – 110 002

ISBN 978-93-5919-566-7

Publisher's Note:

Every possible effort has been made to ensure that the information contained in this book is accurate at the time of going to press, and the publisher and author cannot accept responsibility for any errors or omissions, however caused. No responsibility for loss or damage occasioned to any person acting, or refraining from action, as a result of the material in this publication can be accepted by the editor, the publisher or the author. The Publisher is not associated with any product or vendor mentioned in the book. The contents of this work are intended to further general scientific research, understanding and discussion only. Readers should consult with a specialist where appropriate.

Every effort has been made to trace the owners of copyright material used in this book, if any. The author and the publisher will be grateful for any omission brought to their notice for acknowledgement in the future editions of the book.

Published by : **Daya Publishing House®**
A Division of
Astral International Pvt. Ltd.
– ISO 9001:2015 Certified Company –
4736/23, Ansari Road, Darya Ganj
New Delhi-110 002
Ph. 011-43549197
E-mail: info@astralint.com
Website: www.astralint.com

Printed at : **Replika Press Pvt. Ltd.**

Preface

A recent survey by the Economist revealed that the world population has increased by 90 per cent in the past 40 years while food productions has increase only by 25 per cent head. With an additional 1.5 billion mouth to feed by 2024, farmers worldwide have to produce 39 per cent more. The new millennium promises excitement and hope for the future by new advancement in agriculture research. Doubling farmers' income is the mission of Government of India to achieve the mission. Indian agriculture awaits transformation through innovation technological interventions.

There has been a long felt need for a book exclusively on the "**Agricultural Technologies"** An effort has been made to collect information on various Agricultural Technologies viz Plant Pathology, Entomology, Agri. Economics, Agronomy, Agri. Extension, Fruit Science, Vegetable Science, Veterinary sciences etc. These Technologies are commonly being used by farmers of developing world resulting in reaping a rich harvest. Chapters in this book have been contributed by noted agricultural Scientists of SKUAST Jammu. I am grateful and indebted to all the learned galaxy of contributors who have spent their considerable time in contributing the chapters.

A deep sense of respect and gratitude to Prof. Pardeep K. Sharma, Hon'ble Vice Chancellor, SKUAST-Jammu.

Last but not the least, I must acknowledge and oblige the M/S Astral International Pvt Ltd, New Delhi who has been always considerate and positive to our contentions and efforts.

Dr. Shahid Ahamad

Contents

Division of Plant Pathology

Division of Vegetable Science and Floriculture

Division of Sericulture

Division of Soil Science and Agricultural Chemistry

Division of Agricultural Economics and Agribusiness Management

Division of Food Technology

Division of Agricultural Engineering

AICRP-PRSS, Samba

AICRP-IWM, Chatha

AICRP-Wheat and Barley

Research Stations/Sub Stations: Advanced Centre for Rainfed Agriculture, Rakh Dhiansar, Jammu

Advanced Centre for Horticulture Research, Udheywalla, Jammu

Organic Farming Research Centre, Chatha

RHRSS, Bhaderwah

Regional Agricultural Research Station, Rajouri

School of Biotechnology

Division of Plant Physiology

Krishi Vigyan Kendras

Division of Animal Nutrition

Division of Veterinary Parasitology

Division of Veterinary Medicine

Division of Veterinary Gynecology and Obstetrics

Division of Veterinary Pathology

Division of Veterinary Public Health and Epidemiology

Division of LPM

Division of Veterinary Surgery and Radiology

Division of Livestock Products Technology

Division of VPB

List of Contributors

	Faculty of Agriculture	**Name of Scientist**
	Division of Agronomy	
1.	Crop Diversification for Sustainable Productivity and Higher Land Use Efficiency	*Dr. B. C. Sharma Prof. & Head, Division of Agronomy*
2.	Direct Seeded Rice (DSR) for Efficient Resource Utilization	*Dr. R. Puniya Asstt. Professor, Division of Agronomy*
3.	Zero-till Technology in Wheat	*Dr. B. C. Sharma Prof. & Head, Division of Agronomy*
4.	Rejuvenation of Weed Infested Grazing Lands	*Dr. B. C. Sharma Prof. & Head, Division of Agronomy*
5.	Ring-pit Planting Method of Sugarcane	*Dr. B. R. Bazaya, Sr. Scientist , AICRP-Weed Management*
6.	Utilization of Weed flora Biomass for Vermicomposting	*Dr. B. R. Bazaya, Sr. Scientist , AICRP-Weed Management*
	Division of Entomology	
7.	IPM in Cole Crops, Brinjal and Okra	*Dr. Uma Shankar, Assoc. Professor, Division of Entomology*
8.	Management of Grasshoppers	*Dr. R. M. Bhagat, Ex-Head, Division of Entomology*
9.	Management of *Varroa destructor* Mite	*Dr. D. P. Abrol Ex-Head, Division of Entomology*
10.	Management of pod borer on marigold	*Dr. Hafeez Ahmad Prof. & Head, Division of Entomology*
11.	Management of Wheat Aphid	*Dr. Hafeez Ahmad Prof. & Head, Division of Entomology*
12.	Management of leaf folder in Paddy	*Dr. Hafeez Ahmad Prof. & Head, Division of Entomology*
	Division of Plant Pathology	
13.	Round the Year cultivation of Mushrooms	*Dr. Sachin Gupta Assoc. Prof. Division of Plat Pathology*

	Faculty of Agriculture	**Name of Scientist**
	Division of Vegetable & Floriculture Science	
14.	Integrated nutrient management in vegetable crops	*Dr. Manoj Kumar, Asstt. Professor, Divi. Vegetable Science.*
15.	Portray Nursery Production Technology	*Dr. Sanjeev Kumar, Professor. Asstt. Professor, Divi. Vegetable Science.*
16.	Seed Production of knol-khol under non-traditional environment.	*Dr. Satis Kumar, Assoc. Professor, Divi. Vegetable Science.*
17.	Rapid Multiplication of Tomato Hybrids through Plant Tissue culture	*Dr. Anil Bhusan, Assoc Professor, Divi. Vegetable Science.*
18.	Rapid multiplication of Ginger planting material by Low Cost Protocol	*Dr. R. K. Gupta, Professor & Head Divi. Vegetable Science.*
19.	Rapid Potato multiplication through micro tubers methodology	*Dr. R. K. Gupta, Professor & Head Divi. Vegetable Science.*
20.	River Bed Production of Early Vegetables	*Dr. R. K. Samnotra, Professor Divi. Vegetable Science.*
21.	Knolkhol production Round the Year	*Dr. R. K. Samnotra, Professor Divi. Vegetable Science.*
	Division of Sericulture	
22.	Chawki Rearing Technology for Increasing Cocoon Production	*Dr. R. K. Bali Prof. & Head, Division of Sericulture*
23.	PO3 x ND5: A Productive Bivoltine Hybrid	*Dr. R. K. Bali Prof. & Head, Division of Sericulture*
	Division of Soil Science & Agri. Chemistry	
24.	Soil Fertility Mapping of Jammu district	*Dr. Vikas Sharma, Professor, Division of Soil Science & Agri. Chemistry*
25.	Assessment of physical health and fertility status in soils of Udhampur district of Jammu	*Dr. Piyush Sharma, Assoc. Professor Division of Soil Science & Agri. Chemistry*
26.	Effect of integrated use of biochar and organic manures on soil properties and growth of knolkhol (*brassica oleracea var. gongyloides*)	*Dr. Piyush Sharma, Assoc. Professor Division of Soil Science & Agri. Chemistry*
27.	Impact of biochar application on soil physical properties, soil organic carbon dynamics and crop performance in rice- wheat system	*Dr. Piyush Sharma, Assoc. Professor Division of Soil Science & Agri. Chemistry*

	Faculty of Agriculture	**Name of Scientist**
28.	Effects of phosphorus and sulphur interaction on nutrient uptake and yield of black gram (*vigna mungo* l. Hepper)	*Dr. A. P. Rai. Assoc. Professor*
29.	Response of zinc and iron application in rice (cv. Basmati-370)	*Dr. A. P. Rai. Assoc. Professor*
30.	Management of mustard crop through Sulphur and Zinc application	*Dr. A. P. Rai. Assoc. Professor*
31.	Training and demonstration of LCC for real time nitrogen management under climate smart agriculture practices (CSAP) for maintaining ecological integrity through KVK'S.	*Dr. Vivak M. Arya, Asstt. Prof.*
32.	Assessment of fertilizer recommendations in Maize-wheat, based on Target Yield concept (TYC) under subtropical plains of Jammu.	*Dr. Vivak M. Arya, Asstt. Prof.*
Division of Agricultural Economics and Agribusiness Management		
33.	Economics of Niche crops and assessment of farm level diversification in Jammu region	*Dr. Sudhakar Dwivedi, Associate Professor*
34.	An Economic analysis of Production and Marketing of Rajmash Crop in Doda District of J&K State	*Dr. S. P. Singh, Assistant Professor*
35.	Economic Analysis of Production and Marketing of Marigold Flowers in Jammu Subtropics	*Dr. Anil Bhat, Assistant Professor*
Division of Food Technology		
36.	Development of Fiber Rich Oat-Wheat Flour Blended Biscuits Using Mango Peel	*Dr. Jagmohan Singh, Assoc. Professor*
37.	Effect of pre-treatments and dehydration on the quality of oyster mushroom	*Dr. Anju Bhat, Prof.*
38.	Development and Quality Evaluation of Brown Rice Based Weaning Food	*Dr. Monica Sood, Assoc. Prof.*
Division of Agri. Engineering		
39.	Weeding -Cum- Earthing Up Equipment	*Dr. J. P. Singh, Assoc. Professor*

	Faculty of Agriculture	Name of Scientist
40.	Development and Evaluation of Automatic Timer Based Variable Speed Device for Sprinkler System	*Dr. Sushmita M. Dadhich, Assistant Professor, Division of Agricultural Engineering*
41.	Design and development of a tractor operated soil compaction measurement device	*Er. Hemant Dadhich, Assistant Professor, Division of Agricultural Engineering,*
	Division of Fruit Science	
42.	Characterization of Walnut ((*Juglans regia* L.) Germplasm in Jammu Province	*Dr. Parshant Bakshi, Associate Professor, Division of Fruit Science*
43.	Characterization of Pecan nut (*Carya illinoensis K.*) germplasm of Jammu province	*Dr. Arti Sharma, Asstt. Professor, Division of Fruit Science*
	Division of Agroforestry	
44.	Standardize harvesting time, pre-sowing seed treatment, identification of superior genotype of Harad and its vegetative propagation.	*Dr. K. K. Sood, Professor*
45.	Standardization of agro-technology of medicinal plants: *Aloe barbadensis, Gloriosa superba, Rauvolfia serpentina and Silybum marianum*	*Dr. L. M. Gupta, Professor*
	AICRP-IFS	
46.	Diversification/ Intensification: An alternative cropping system for enhancing Farm Productivity and Employment under irrigated conditions of Jammu	*Dr. Dileep Kachroo, Chief Scientist ,*
47.	Integrated Farming System Model towards round the year production, income and employment generation for small and marginal farmers.	*Dr. Dileep Kachroo, Chief Scientist ,*
	AICRP- Rice	
48.	System of Rice Intensification – efficient input use methodology	*Dr. Anuradha Saha, Sr. Scientist (Agronomy)*
	AICRP- Weed Management	
49.	Weedy Rice Management in Transplanted Rice	*Dr. B. R. Bajaya, Assoc. Prof*

	Faculty of Agriculture	**Name of Scientist**
AICRP- Rapeseed-Mustard		
50.	Moisture stress mitigation in Indian mustard using agrochemicals	*Dr. Rajeev Bharat, Jr. Scientist (Agronomy)*
AICRP-PRSS, Samba		
51.	Bio-efficacy of different herbicides for broad spectrum weed management in chickpea Contact Scientist:	*Dr Brij Nandan , Senior Scientist (Agronomy)*
AICRP-IWM, Chatha		
52.	Laser leveling for improving yield and saving water in rice-wheat sequence under irrigated conditions of Jammu	*Dr. Vijay Bharti, Assoc. Prof.*
53.	Multiple water use model for Waterlogged Areas of Jammu region	*Dr. Vijay Bharti, Assoc. Prof.*
54.	Increased Potato Production and Water Saving through Sprinkler Irrigation in Canal Command of Jammu	*Dr. Vijay Bharti, Assoc. Prof.*
AICRP-Wheat		
55.	Management of powdery mildew diseases of wheat	*Dr. M. K. Pandey, Sr. Scientist*
Research Stations : ACRA, Rakh Dhiansar		
56.	Integrated nutrient management in rainfed maize	*Dr. Vikas Abrol, Assoc. Prof*
57.	Biopesticidal management of mustard aphid	*Dr. Reena, Sr. Scientist*
58.	Mustard aphid management through organic and inorganic plant nutrition	*Dr. Reena, Sr. Scientist*
ACHR, Udheywalla, Jammu		
59.	Bio-control of soil- borne diseases of vegetables crops	*Dr. Vishal Gupta, Asstt. Professor*
OFRC, Chatha		
60.	NSKE: A Potent Bio-Pesticide for Insect pest and diseases management	*Dr. S. K. Singh, Sr. Scientist, Plant Pathology*
61.	High Density Organic Guava Cultivation	*Dr. Akas Sharma, Asstt. Professo*

	Faculty of Agriculture	Name of Scientist
	RHRSS, Bhaderwah	
62.	Effect of nutrient management in maize+rajmash intercropping in sub-alpine conditions of Bhaderwah.	*Dr. Rohit Sharma, Asstt. Professor*
63.	Agrotechnique Refinement for wild pomegranate (*Anardana*) in J & K.	*Dr. P. Bakshi, Sr. Scientist*
	RARS, Rajouri	
64.	Effect of Row Ratio of Parents and Spacing on Single Cross Hybrid Seed Production of Quality Protein Maize (*Zea mays* L.)	*Dr. Vikas Sharma , Sr. Scientist*
65.	Green fodder supply round the year in Rajouri district	*Dr. Vikas Sharma , Sr. Scientist*
66.	Rainwater harvesting and its demonstration for supplemental benefits for high value crops in hill region.	*Dr. Vikas Sharma , Sr. Scientist*
	B. School of Biotechnology	
67.	Grain quality analysis and molecular characterization (DNA fingerprinting) of basmati rice	*Dr. Romesh K Salgotra Professor*
68.	Molecular characterization: A tool for diversity analysis and varietal identification in vegetable crops using DNA markers	*Dr. Susheel Sharma, Asstt. Professor,*
	Faculty of Basic Science, Chatha	
	Division of Plant Physiology	
69.	Management of un-fruitfulness in Olive	*Dr. A. K. Tiku. Ex-Head*
	D. Krishi Vigyan Kendras	
70.	Mitigation of stripe rust (yellow rust) disease of wheat through fungicides and varieties.	*Dr. Shahid Ahamad, Chief Scientist & Head, Krisihi Vigyan Kendra, Reasi*
71.	Rhizome rot of ginger pose a threat to ginger production hill region	*Dr. Shahid Ahamad, Chief Scientist & Head, Krisihi Vigyan Kendra, Reasi*
72.	Organic management of rhizome rot of ginger .	*Dr. Shahid Ahamad, Chief Scientist & Head, Krisihi Vigyan Kendra, Reasi*

	Faculty of Agriculture	**Name of Scientist**
E. Faculty of Veterinary Sciences & AH, R S Pura, Jammu		
Division of Animal Nutrition		
73.	Condensed tannins enriched multi-nutrient blocks For Livestock Productivity Enhancement	*Dr. R. K. Sharma, Prof. &Head.*
Division of Veterinary Parasitology		
74.	Assessment of economic losses caused by *Przhevalskiana silenus* infestation in goats of Jammu region and development of Cost effective treatment regimen	*Dr. Rajesh Katoch, Prof. & Head.*
75.	Detection of acaricide resistance in ticks	*Dr. Rajesh Katoch, Prof. & Head.*
Division of Veterinary Medicine		
76.	Establishment of early diagnostic markers for the diagnosis of bovine mastitis	*Dr. Neelesh Sharma Asstt. Prof.*
77.	Development of cost effective technology for the management of mastitis in dairy animals	*Dr. Neelesh Sharma Asstt. Prof.*
Div. of VGO, FVSc & AH		
78.	Embryo transfer technology in sheep	*Dr. Anil Kumar Pandey, Assistant Professor*
79.	Soya based extender and use of various semen additives for cryo preservation of cattle and buffalo semen	*Dr. W A A Razzque, Asstt. Prof,*
Division of Veterinary Pathology		
80.	Molecular characterization and phylogenetic analysis of different avipox viruses	*Dr. Nawab Nashiruddullah, Professor & Head*
81.	AgNOR as a marker of cell proliferation in differentiating benign and malignant tumors of domestic animals	*Dr. Shafiqur Rahman, Assistant Pofessor,*
82.	Etio-pathomorphological studies on gastro-intestinal tract of broiler chickens in Jammu.	*Dr. Shilpa Sood, Associate Professor*

	Faculty of Agriculture	**Name of Scientist**
	Division of Veterinary Public Health & Epidemiology	
83.	Scientific interventions for control of zoonotic parasites of Livestock and livestock rearers	*Dr. Mohd. Rashid Assistant Pofessor,*
84.	Public Health Significance of Echinococcosis in Jammu	*Dr. M. A. Malik, Prof. &Head.*
85.	Enterotoxigenic profile of *Bacillus cereus* in meat and meat products	*Dr. M. A. Malik, Prof. &Head.*
	Division of LPM	
86.	Heat stress amelioration in dairy cattle by showering.	*Dr. Asma Khan, Prof. & Head*
87.	Pine leaves as bedding material improves economics and performance of broiler chicks.	*Dr. B. Brahma, Assoc. Prof.*
88.	Enhancement of growth performance and meat quality of *Cyprinuscarpio* by chicken manure.	*Dr. Sahar Masood, Assoc. Professor*
89.	Turkey production under intensive and semi-intensive rearing system.	*Dr. Asma Khan, Prof. & Head*
	Division of VSR	
90.	Balanced total intravenous Anaesthesia protocols for adult mules	*Dr. MS Bhadwal, Professor*
91.	Management of Intussusception in cattle	*Dr. H. R. Bharaddwaj, Asstt. Professor*
92.	Post operative pain management in dogs	*Dr. M. S. Bhadwal, Professor*
93.	Ultrasound guided and laparoscopic needle biopsy in dogs	*Dr. A. K. Gupta, Professor*
94.	Ultrasound guided biopsy of liver and kidney in cattle.	*Dr. M. S. Bhadwal, Professor*
	Division of livestock Products Technology	
95.	Development of fiber enriched chicken sausages	*Dr. Sunil Kumar, Professor*
96.	Shelf-stable fish snacks	*Dr. Sunil Kumar, Professor*
97.	Designer fiber-enriched chevon cutlets	*Dr. Sunil Kumar, Professor*

	Faculty of Agriculture	**Name of Scientist**
98.	Development of composite meat chocolate	*Dr. Sunil Kumar, Professor*
99.	Fruit fiber fortified emulsion based spent hen meat products (patties)	*Dr. Sunil Kumar, Professor*
	Division of VPB	
100.	Effect of Seabuckthorn (*Hippophae rhamnoides)* Leaf Extract on biochemical and antioxidant alterations in experimentally induced diabetes mellitus and lead toxicity in Wistar rats.	*Dr. Aditi Lal Koul, Asstt. Professor*

Faculty of Agriculture

Division of Agronomy

1. Crop Diversification for Sustainable Productivity and Higher Land Use Efficiency

Contact Scientist: **Dr. B.C. Sharma**

Rice–Wheat is the existing popular cropping system of irrigated sub-tropics of Jammu and Kashmir. Since both the rice and wheat crops are exhaustive in nature and continuous cultivation of these crops for longer periods has led to depletion of the soil nutrients and this has resulted in reduction of potential productivity of these crops. In order to restore the soil fertility and ensure sustainable crop productivity, there was a need to identify need based sustainable and economic cropping systems as an alternative to the existing rice-wheat system.

View of the Different Field Activities and Crops under Taken for Experimentation.

Methodology

The crops selected for different diversified cropping systems were raised with their respective recommended packages of cultivation. The inclusion of the crops in different cropping systems have been made keeping in view the duration of the crops.

Table 1: Rice-Jaya Equivalent Yield, Land Use Efficiency, Production Efficiency, Employment Generation and Economics of different Crop Sequences

Crop Sequence	*Rice-Jaya Equivalent Yield (q/ha)*	*Lane Use Efficiency (per cent)*	*Production Efficiency (kg/ha/ day)*	*Employment Generation (Man-days/ year)*	*Net Returns (Rs/ha)*
Rice (Jaya)-Wheat (PBW-343)-Fallow (Check)	124.53	70.90	48.08	163	30012.40
Rice(IET-1410)-Berseem (Vardan) for fodder and seed production	133.39	88.70	41.17	215	35477.20
Rice (PC-19)-Wheat (PBW-343)-Fodder (local maize+ cowpea+ charri)	150.62	85.40	48.27	212	34773.60
Rice (IET-1410)-Potato (Kufri badshah)-Wheat (PBW-373)	265.57	85.75	84.84	280	60829.30

Recommendations

Identified need based, productive, economical and sustainable alternatives of 300 per cent cropping intensity for the diversification of existing rice-wheat cropping system in irrigated sub-tropics of Jammu and Kashmir with a net return of Rs. ranging from 30,000-61,000/ha for different cropping systems.

Benefits/Impacts

Adoption of identified sequences may not only enhance the productivity and net returns but will also improve the soil health as these are quite restorative as compared to existing cereal-cereal based cropping sequence.

2. Direct Seeded Rice (DSR) Technology

Contact Scientist: Dr. R. Puniya

Direct seeding of rice is the process of establishing the crop from seeds rather than by transplanting of seedlings. Direct seeding avoids three basic operations, namely, nursery raising, puddling and transplanting.

Methodology

- Ploughing with SS plough followed 2-3 harrowings and planking under optimum moisture (Vattar conditions).
- Sowing of seeds 10- 15 days prior to onset of monsoon (5th June to 15th June)
- 30 and 40 seed kg/ha for fine and coarse varieties, respectively.
- Treat the seeds with Bavistin @ 2 g/kg of seed.
- Seeds should be sown at specified spacing of 20 cm by seed cum fertilizer drill at a depth of 3-4 cm.
- For proper weed management apply pendimethalin 1 L/ha (PE) in 500 liters followed by bispyribac sodium @ 25-30 g/ha at 25-30 DAS using in 500 liters of water for each herbicide.

Table 1: Comparative Economics of Conventional and Direct Seeded Rice (Jaya) Grown at KVK, Kathua

Cultivation Techniques	*Yield/ha*	*Cost of Cultivation (Rs./ha)*	*Gross Returns (Rs./ha)*	*Net Returns (Rs./ha)*	*B:C Ratio (Re/Re)*
Conventional Rice	42.12	21500	71750	50250	2.01
Direct Seeded Rice	40.23	18250	68960	50710	2.77

Table 2: Comparative Economics of Conventional and Direct Seeded Rice (Basmati-370) grown at Research Farm of Agronomy at Chatha

Cultivation Techniques	*Yield/ha*	*Cost of Cultivation (Rs./ha)*	*Gross Returns (Rs./ha)*	*Net Returns (Rs./ha)*	*B:C Ratio (Re/Re)*
Conventional Rice	29.0	32,000	116000	84000	2.63
Direct Seeded Rice	28.5	25,000	114000	89000	3.56

Field View of Direct Seeded Rice (Jaya) at KVK, Kathua.

Field View of Direct Seeded Rice (Basmati- 370) at Research Farm of Agronomy.

Recommendation

In order to facilitate timely planting of rice in present day scenario of labour constraints, direct seeding of rice is only the viable option as in this technique the yield results are by and large not only statistically at par with conventional rice but has also been found superior in terms of net returns and benefit cost ratio.

Benefits/Impacts

1. Facilitates timely planting of rice under labour scarcity
2. Saving of water
3. Matures 7-10 days early and thus helps in timely sowing of succeeding crops.
4. Comparatively eco-friendly.
5. Comparatively less disturbance of soil structure
6. Yields are at par with transplanted rice but it has higher comparatively net saving of Rs. 5000-6000/ha.

3. Zero-till Technology in Wheat

Contact Scientist: Dr. B.C. Sharma

Zero tillage is an extreme form of minimum tillage where in primary tillage is completely avoided and secondary tillage is restricted to seedbed preparation in the row zone only. Zero till machine accomplishes three tasks in one operation *i.e.* opening of narrow slits for insertion of seed, fertilizer and their coverage.

Methodology

- ❒ In weed infested fields spray glyphosate @ 1 L/ha in 500 L of water.
- ❒ Use 120-130 kg seed/ha
- ❒ The seeds should be sown at specified row spacing's of 20 cm width with the help of tractor drawn zero till drill at 4-5 cm depth.
- ❒ For weed management apply suflosulfuron + metsulfuron 30+ 2 g/ha or clodinafop + metsulfuron 60 g/ha at 30-35 DAS (post-emergence) of wheat crop in 500 L of water.

Table 1: Effect of Tillage Systems on Grain Yield and Economics of Wheat Grown at Research Farm of Agronomy, Chatha

Tillage Systems	*Grain Tield/ha*	*B:C Ratio (Re/Re)*
Zero-tillage	42.44	2.12
Conventional tillage	40.96	1.55

Zero seed cum ferti-drill in operation

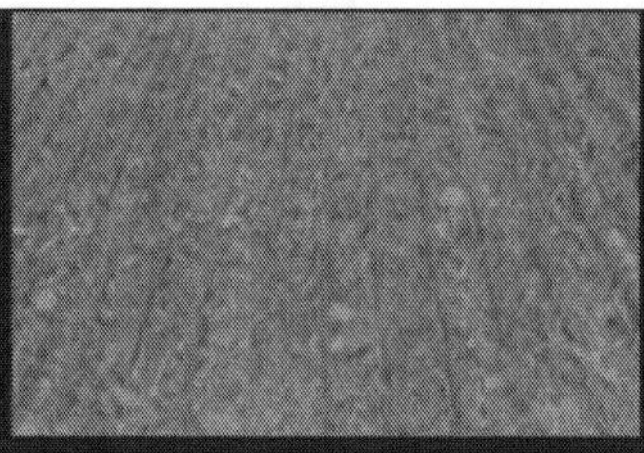

Zero till wheat crop at heading

Zero till established wheat crop

Recommendations

In order to facilitate timely sowing of wheat specifically under Basmati rice-wheat system Zero-till sowing is the better option as in this technique the yield results are by and large not only statistically at par with conventionally sown wheat but has also been found superior in term of net returns and benefit cost ratio.

Table 2: Effect of Tillage Systems on Grain Yield and Economics of Wheat Grown at KVK, Kathua

Tillage Systems	*Yield (q/ha)*	*Cost of Cultivation (Rs./ha)*	*Gross Returns (Rs./ha)*	*Net Returns (Rs./ha)*	*B: C Ratio (Re/Re.)*
Zero tillage	42.54	22550	68064	45514	3.02
Conventional tillage	43.27	26550	69232	42682	2.61

Field View of Zero-Tillage Wheat grown at KVK, Kathua.

Benefits/Impact of Zero Tillage

1. Facilitates timely planting of normal sown wheat
2. Yields are at par with conventional normal sown wheat and comparatively higher by about 3-5 q/ha from late sown conventional wheat
3. Ecologically safe as it saves energy in shape of fuel and reduces cost of cultivation by about Rs. 3000/ha.
4. Increases rate of infiltration, organic matter retention and re-cycling of nutrients in the soil

4. Rejuvenation of Weed Infested Grazing Lands

Contact Scientist: Dr. B.C. Sharma

Lantana camara is an alien invasive weed from South and Central America which was widely introduced as an ornamental plant and is now considered a weed in more than 50 countries worldwide. It has invaded all most all parts of India mainly in intermediate and sub-tropical areas including Jammu region. It forms dense thickets that displace natural communities and compromise carrying capacity of pasture lands. Lantana is a highly variable species, with hundreds of different cultivars that differ in appearance and in their tolerance to environmental conditions.

Species of Lantana with different Flower Colours.

Methodology

1. Cut the Lantana bushes near the ground level in the month of April.
2. Apply glyphosate @ 1 per cent as foliar spray on regenerated growth of about 30 cm.
3. Revegetate the treated area either with Napier or *Setaria* or *Cenchrus* species to prohibit its reoccurrence.

Site view of *Saccharum*, *Parthenium* and *Lantana* infested land before treatments

View of Weeds after herbicidal treatments

Rejuvenated weed infested land after experimentation

Recommendation

Successful rejuvenation of *Saccharum spontaneum, Parthenium hysterophorus* and *Lantana camera* infested grazing lands can be achieved by integrating management options like cutting of lantana bushes followed by application of glyphosate @ 1 per cent as foliar spray on regenerated growth of about 30 cm. The treated land is provided soil cover by planting of perennial grasses (Napier, *Setaria, Cenchrus*) to avoid further encroachment by the weeds.

Benefits/Impact

Though there is no authentic statistics with regards to infestation of these obnoxious weeds in grazing/pasture lands in our state but if we go by the visual observations most of the uncultivated as well as other common/shamlat lands falling in the altitudinal range of 800-3000 amsl are likely to prone to the menace of these weeds. A management of such degraded lands may provide additional fodder which may reduce the dependence of the state on import of fodders from other states.

5. Ring-pit Planting Method of Sugarcane

Contact Scientist: Dr. B.R. Bazaya

In Jammu and Kashmir sugarcane crop occupies an area of 1.93 thousand hectares with production of 3000 quintals. The productivity of sugarcane in Jammu region is very low than the national average. Jammu and Kashmir State is importing sugar from other states causing huge loss to financial exchequer of the State. In general, flat planting of sugarcane is extensively followed in Jammu region and this technique of planting debars to realize the potential yield of sugarcane due to poor tillering and lodging of crop. A new method of sugarcane cultivation called as ring-pit method, is not only cost-effective but also helps in attaining higher yield.

Methodology

- Circular of pits of 90 cm circumference and 45 cm depth are to be dug at spacings of 60 × 60 cm. At this spacing about 4,000 pits can be dug per hectare. The pits are to be refilled with a mixture of soil and FYM up to 15 cm depth.
- Plant 16-22 setts (two budded) or 30-40 setts (single budded) per pit horizontally in cyclic manner and before planting dip the setts in 0.1 per cent solution of carbendazim for 30 minutes for control of diseases. The can thus planted are covered with soil to a thickness of 5-7 cm.
- Fertilizer dose: N:P:K:: 150:75:75 kg/ha (Urea : 264 kg, DAP : 165 kg, MOP 125 kg/ha.). Apply ½ N and total P and K at planting time and remaining N in two equal splits *i.e.* 1/4th in the month of June and the rest 1/4th at the end of grand growth phase in August.

Table 1: Cane Yield under Ring-Pit and Flat Bed Methods of Planting

Planting Methods	*Cane Yield (q/ha)*	*Net Return (Rs/ha)*	*B:C Ratio*	*Extra Yield Advantage over Flat Bed Method (q/ha)*
Ring-pit method	2061	213518	3.15	123
Flat bed method	1938	199419	2.99	-

Recommendation

Ring-pit method of sugarcane planting has been found most suitable as it helps in enhancing cane yield, net returns and benefit cost ratio over traditional flat bed method under irrigated conditions of Jammu region.

Ring-Pit Planting Method of Sugarcane.

Benefits/Impact

- Ring-pit method gives higher extra cane yield (123 q/ha) than traditional flat bed method.
- Prevents lodging
- Sugarcane planted with ring-pit method can provide 1-2 ratoons without much reduction in cane yield.
- Water use efficiency is increased by 30-40 per cent and nutrient use efficiency by 30-35 per cent due to their localized application in pits only.

6. Utilization of Weed Flora Biomass for Vermicomposting

Contact Scientist: Dr. B.R. Bazaya

Vermicomposting is a process in which earthworms are used to convert organic materials into humuslike material known as vermicompost. Vermicompost is nutritive organic manure rich in NKP, micronutrients, and beneficial microbes. Weeds are unwanted plants but some of them can be converted into valuable source of nutrients after its conversion to compost through the process of vercomposting. Farmers can be produce vermicompost from the weeds, which are mostly available on their own farms and they need not spend extra money to purchase the raw material of required for vermicompost.

Methodology

- Three major obnoxious weed species having high nutritive value selected for vermicomposting *viz., Eichhornia crassipes Lantana camara* and *Leucaena spp.*
- The biomass of weed flora mixed before flowering with FYM in 1:2 combinations.
- Fresh dung should be avoided. FYM should be used 10-12 days old at the time of pit filling.
- Weed flora leaves should be chopped in small pieces before mixing with worms in pits.
- Weed flora should be mixed before the flowering of weeds and avoided after seed setting.
- Place the chopped weed biomass and partially decomposed FYM layer-wise in the pit.
- Maintain 50-60 per cent moisture content as per need through water spray in vermi-pit.

Table 1 Effect of different Weeds Flora Biomass on Worm Populations, Compost Production and Nutrient Contents

Treatments	*Worm Population (no/m³)*	*Eggs (no/m³)*	*Vermicompost Production (kg/m³)*	*Nutrient Content (per cent)*		
				N	*P*	*K*
Leucaena spp.+ FYM	1738.0	400.7	57.04	2.42	0.87	1.90
Lantana camara + FYM	1466.2	267.8	53.98	2.34	0.97	1.84

Treatments	Worm Population (no/m^3)	Eggs (no/m^3)	Vermicompost Production (kg/m^3)	Nutrient Content (per cent)		
				N	P	K
Eichhornia crassipes + FYM	1797.6	404.8	58.23	1.92	0.88	1.91
FYM alone	1327.3	166.6	53.07	1.52	0.68	1.49

Eggs of Earthworms

Earthworms

View of the different Activities of Vermicomposting.

Recommendation

Leucaena + FYM (1:2) or *Eichhornia* + FYM or Lantana+ FYM have been adjudged as the most suitable weed flora biomass for vermicomposting.

In order to attain higher nutrient content of major plant nutrients in the vermicompost, following combinations of weed flora biomass and FYM are recommended

1. For higher nitrogen content *Leucaena* + FYM (1:2)
2. For higher phosphorus content *Lantana* + FYM (1:2)
3. For higher potassium content *Eichhornia* + FYM (1:2)

Benefits/Impact

- Vermicomposting is an alternate approach for recycling of weed flora. It adds essential plant nutrients and organic matter in soil and is also cost effective.
- Improves soil health and sustain it for long period as it is eco-friendly
- Improve quality of produce and crop yields.
- Can be adopted an entrepreneur unit by unemployed rural youths of the region as it will generate income through vermicomposting.

Division of Entomology

7. IPM in Cole Crops, Brinjal and Okra

Contact Scientist: Dr. Uma Shankar

Among the various important vegetables that are produced in Jammu and Kashmir, the vegetables *viz.*, tomato, cauliflower, cabbage, broccoli and Knol-khol, okra, brinjal and cucurbits are the important ones that add higher revenue to the state. The major constraints in the production of these crops is the ravages caused by various insect pests and diseases which not only result in the low production of vegetables but also drastically impair the quality of these, rendering them unfit for human consumption, reducing marketable yield and thereby posing a serious threat to farming communities.

IPM is the worldwide accepted technology to reduce the pesticide pressure on the crops. Therefore, development of regional IPM technology is one of the possible strategies to manage pest problems and thereby boosting the yields of vegetable crops. Integration of Various IPM tools include use of various bioagents, pheromones, light traps, biological formulations (Bt and NPV), botanical formulations (NSKE or neem oil formulation, pongamia soap formulation *etc.*), inter-cropping (either as trap crop or as deterrent crop) and also management of nursery by soil solarization, seed treatment with bioagents like *Trichoderma etc.* and covering the nursery beds with agro-nets are the strategies were vFFalidated on the fields of farmers in Jammu region. Selective eco-friendly chemicals were also used for managing the pest populations as a last resort to save their produce.

Methodology

Various regional specific IPM modules including IPM devices and microbial pesticides were formulated at different crop growth stages to mitigate the losses caused by insect pests and diseases in nursery raising as well as in field conditions on cole crops, tomato and brinjal. Various eco-friendly chemicals and bio-pesticide were also evaluated for managing the pest population on these crops.

Table 1: Reduction in Pesticide Spray and Yield Enhancement in Selected Vegetable Crops at Jammu Region

Crops	*Pesticides Pressure in Non-IPM field (Spray)*	*Pesticides Reduction in IPM Intervened Field (Spray)*	*Yields Enhancement in (per cent)*
Cole crops	12	5	15
Tomato	12	3-4	30
Brinjal	25-30	6-8	22-30
Okra	10-12	4-5	10-20

Nursery Raising at Basht

Low Cost Polyhouse Brinjal Seedlings at Marh Block

Tricho-card in Cauliflower Field

Pheromone Trap to catch *Spodoptera*

Field Demonstration to Farmers

Traps in Tomato Field Pahariwala

Recommendations

Nursery Management in Vegetable Crops

- Raised nursery beds 8 cm above normal ground level helped in growing better seedlings in nursery.
- Soil solarization initiated in the month of June for 6-8 weeks in conjugation of well rotten FYM and *Trichoderma* powder (*T. viridae* and *T. harzianum*) helped in growing disease free seedlings in nursery for early sowing.
- Virus free tomato seedlings were raised using Nylon netting.
- Root dip of tomato seedling for 30 minutes in imidacloprid @ 0.3 ml per lit of water helped in avoiding virus transmission.

IPM Technology for Cole Crops

- Hand picking and mechanical destruction of egg masses and caterpillars especially for cabbage butterfly and *Spodoptera litura* was found to be promising
- Release of *Trichogramma chilonis* @ 50,000 adults/ha per release (6 times) at weekly interval in Mid January helped in reducing damage
- Planting border row of mustard and coriander as flowering plants encouraged the *Cotesia glomeratus,* potential parasitoids against cabbage butterfly larva.

- Biopesticide like Neem oil (4 per cent) and NSKE 4 per cent as a blanket spray minimized the hatching of caterpillars in cole crops
- Foliar spray of *Bacillus thuringiensis* formulations @ 500 g/ha along with sticker (0.5 ml/lit of water) or spinosad @ 0.5 ml per lit or cartap hydrochloride @ 1 ml per lit of water of was found promising to control all lepidopteran insect pests in cole crops.

IPM Technology for Tomato

- Broadcasting carbofuron 3G @ 1 kg per kanal at the time of field preparation was found effective in the mitigating the hibernating insect pests.
- With the initiation of flowering (30 days after transplantation) installation of pheromone traps @ 5-7 per ha for early detection and 12-15 per ha was found promising for mass trapping and mating disruption.
- Planting two border row of African marigold provided enough refugia for trapping the adults of *Helicoverpa* for egg laying and also encouraging the larval parasitoid *Camoletes chloridae* for parasitization.
- Spray *Ha*-NPV 300 to 500 LE/ha (freshly prepared) mixed with robin blue and stickers @ 2-3 times at 10 days interval at evening hours was found promising in reducing the larval infestation
- Spraying of bacterial formulation *Bacillus thuringiensis* @ 500 g/ha along with sticker (0.5 ml/lit of water) gave the satisfactory control for *Helicoverpa* larvae in the field condition.

IPM Technology for Brinjal Shoot and Fruit Borer (BSFB) on Brinjal

- Regular clipping of affected twigs and fruits (Once in 10 days) along with the insect helped in reducing the damage
- Broadcast carbofuron 3G @ 1 kg + 1kg dry sand or soil per kanal was found effective for destroying the hibernating insect pests.
- Installation of Wota trap or water trap with leucin lures @ 100 per ha at 10 m spacing trapped the adults moths for mating disruption.
- Use flowering plants in 5-10 per cent of total cropped area for promotion of natural enemy fauna like parasitoids *Trathala flavor-orbitalis, Gonizus* sp. and pradators such as spined pentatomid bug, rediviid bugs and spiders.
- Spraying the crops alternatively with profenophos 50 EC @ 2ml/litre of water at 15 days interval and cartap hydrochloride (1ml/litre of water) starting from 30 days after transplantion was found effective in reducing the brinjal shoot and fruit borer (BSFB) damage.

IPM Technology for Okra (Sucking insect pests)

- Avoid the excessive use of fertilizer doses as it leads to more insect infestation.
- Seed treatment of okra with imidacloprid (Gaucho 600 FS) @ 7 ml (g)/ kg seed provided the protection against sucking pests up to 60 days and also checked the transmission of yellow vein mosaic virus by vectors.
- Two to three spray of botanicals like NSKE 5 per cent solution at 15 days interval gave good control against the pest population.
- Installation of Delta traps or yellow sticky traps to trap the aphids, whitefly *etc.*
- Planting border rows of flowering plants like saunf or coriander or asters encouraged the natural enemy like *Coccinella* (lady-bird beetles), spiders, wasp, green lacewing, rove beetle and syrphid fly, robber fly and other hymenopteran parasitoids for reducing the insect pest population.
- Spraying of imidacloprid @ 0.3 ml or thiomethoxam @ 0.3 ml per lit of water was found promising in case of severe infestation of sucking pests

Borer Insect-Pest Complex on Okra

- Regular clipping off borer infested shoots and fruits helped in reducing the damage.
- Installation of pheromone traps/wota traps @ 100 per ha was found effective in controlling the fruit borer adult moths in okra field.
- Spray NSKE 5 per cent solution @ 3 ml per lit of water followed by cypermethrin @ 1 ml per lit of water or flubendiamide @ 0.3 ml per lit of water was found effective.

Innovative Techniques

Modifications in Pheromone Traps

- Innovative technology has been made to modify the existing pheromone/ sleeve traps by using 2 litre empty bottles of Coca-Cola fixed in existing conical plastic funnel with transparent tape to avoid damage to pheromone/sleeve traps by different birds such as field pipet, Bushchats, Pied bushchats, Black drongo, Grey and black francolins and Common Mynah.
- It was found that modified traps were most successful in catching the moths and without incurring any additional cost.

Impact Analysis/Benefits

The benefits of the IPM technology are as follows:

- IPM modules have resulted in 20-30 per cent increase in yield. The pesticides sprays have come down from twelve to three (need based) in case of tomato.
- Pesticidal treatments have also drastically reduced to 6-8 from 25-30 in brinjal crop.
- Farmers are advised to use various types of pheromone traps, microbial pesticides, eco-friendly pesticides and IPM modules for increased productivity of vegetable crops and reduction in use of pesticides.

8. Management of Grasshoppers

Contact Scientist: Dr. R.M. Bhagat

Methodology

Field trials were conducted to evaluate three different modules for the management of grasshopper. The experiments were carried out at Research farm, Chatha, SKUAST-Jammu. The land preparation, seedling rate, fertilizer doses, irrigation and other cultural operations were performed as per package of practices.

Day before Transplanting	*21 DAT*	*42 DAT*	*63 DAT*
Seed treatment with imidacloprid 30.5 per cent SC @ 5 g kg^{-1} seed at nursery sowing. Dusting on bunds.	Chloropyriphos dust 1.5 per cent @ 10 kg ha^{-1}	Cartap hydrochloride 4G @ 20 kg ha^{-1}	Foliar application of diflubenzuron 25 per cent WP @ 600 g ha^{-1}

DAT: Day after treatment.

Damage Symptoms Paddy

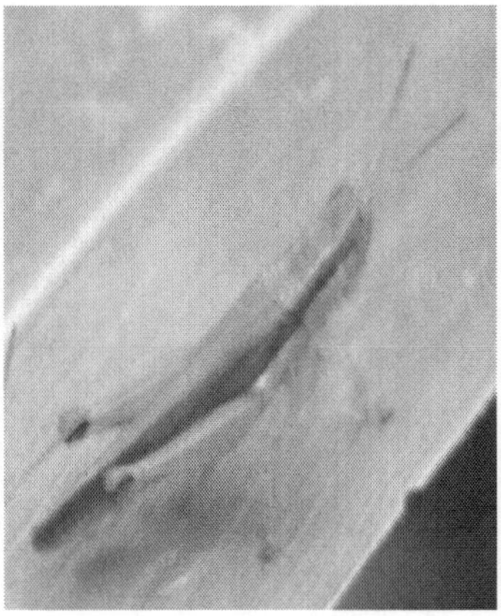

***Oxya velox* Fabricius**

Acrida exaltata

Recommendations

The technology developed proved effective in managing grasshopper menace in paddy growing areas of Jammu region.

Benefits/Impacts

The module developed suppressed the pest populations to economically non threatening levels. Sustainable strategies to minimize the likelihood and extent of grasshopper outbreaks while limiting the need for chemical intervention are a rational and attainable goal for managing grasslands as renewable resources.

Table 1: Evaluation of IPM Module for the Management of Grasshopper

No. of Hopper Quadrat^{-1}											
PTC (20 DAT)	*(22 DAT)*	*(28 DAT)*	*(35 DAT)*	*PTC (41 DAT)*	*(43 DAT)*	*(49 DAT)*	*(56 DAT)*	*PTC (62 DAT)*	*(64 DAT)*	*(70 DAT)*	*(77 DAT)*
Chloropyriphos Dust 1.5 per cent at 21 DAT @ 10 kg ha^{-1}				*Cartap Hydrochloride 4G was Broadcasted in Standing Crop at 42 DAT @ 20 kg ha^{-1}*				*Foliar Application of Diflubenzunon 25 per cent WP at 63 DAT @ 600gm ha^{-1}*			
21.500 ± 0.725	0.400 ± 0.579	17.500 ± 0.851	18.000 ± 0.822	19.100 ± 0.765	16.000 ± 0.316	16.300 ± 1.991	15.400 ± 1.279	16.600 ± 0.731	16.400 ± 1.623	12.700 ± 0.903	8.200 ± 0.515

9. Management of *Varroa destructor* Mite

Contact Scientist: Dr. D.P. Abrol

Methodology

IPM module for management of *Varroa destructor* was developed and tested in different seasons depending upon the brood rearing condition of the colonies. Three applications of each treatment were given in the beginning of each trimester of treatment modules. Pre-treatment count of mites was taken before each application. Thereafter, observations were made after 7, 14 and 21 days after treatment (DAT). For this purpose, 100 cells containing brood were examined for presence of mites.

Results

The efficacy of treatments during different seasons was in the order:

February-April	Queen caging + Cinnamon oil
May-July	Sulphur + oxalic acid
August-October	Formic acid + Cinnamon oil
November-January	Thymol + Oxalic acid

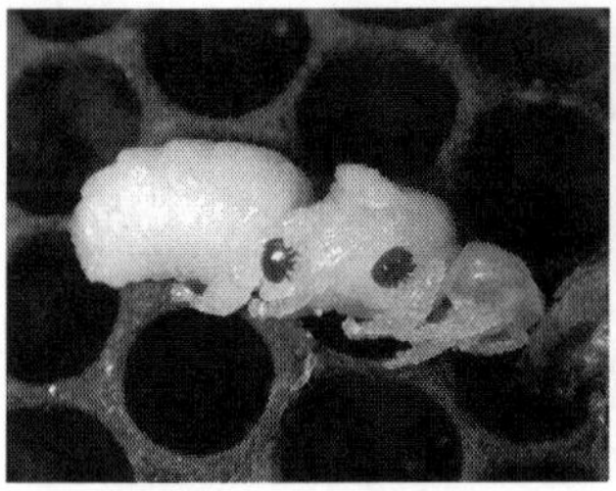

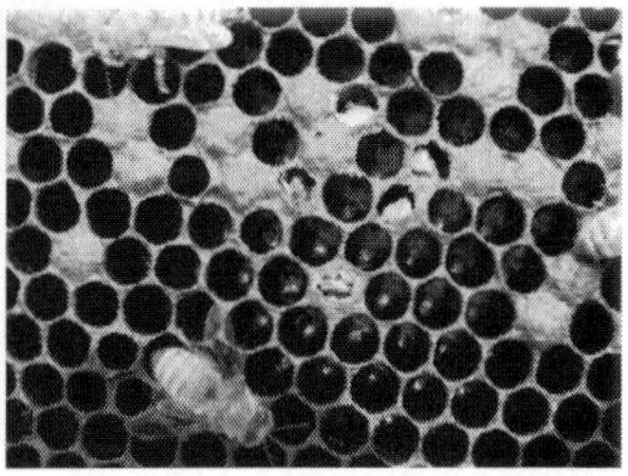

Recommendations

The technology is recommended for managing mites in bee colonies

Benefit/Impact

The *V. destructor* cause severe losses in J&K Punjab, Himachal Pradesh, Haryana and Rajasthan which could also be due to this species. More than 7500 colonies of *A. mellifera* from different apiaries have been sampled for the presence of associated pathogens, mites *etc.* Survey reports have revealed a loss of more than 60 to 100 per cent in some apiaries. The losses in terms of fruit and field crops, from honey, beeswax is un-estimated. The technology developed proved to be a game changer in managing *V. destructor.* The number of bee colonies since then has been increased and at present more than 50,000 bee colonies are there in and K.

10. Management of Pod Borer on Marigold

***Contact Scientist:* Dr. Hafeez Ahmad**

Methodology

Fields trials were conducted to evaluate ten different insecticides including control for the management of pod borer. The experiments were carried out at Research Farm, Chatha, SKUAST-Jammu. The land preparation, seedling rate, fertilizer doses, irrigation and other cultural operations were performed as per package of practices of SKUAST-Jammu.

Recommendations

The technology developed proved effective in managing pod borer menace in marigold grown in Jammu region using carbosulfan followed by novaluron and bifenthrin.

Benefits/Impacts

The farmers in this area are having more than ten years of experience in marigold cultivation and in early days they faced major problem of borers resulting in drastic reduction of flower yield. Because of their lack of knowledge and awareness about modern management practices and inefficient and indiscriminate use of inputs they have faced the problem of reduction in productivity of marigold The recommendation for the *Helicoverpa* management had a major pragmatic impact resulting in contributing to 33-45 per cent higher net profits for smallholder farmers, 25-30 per cent higher yields in IPM plots.

Table 1: Efficacy of different Insecticides against Pod Borer on Marigold

Treatments	*Concen-tration/ Dose*	*Mean Helicoverpa Population/Plant*								*Yield Over Control q/ha*
		1st spray				*2nd Spray*				
		1 DBS	*1 DAS*	*7 DAS*	*14 DAS*	*1 DBS*	*1 DAS*	*7 DAS*	*14 DAS*	
Novaluron 10 EC	**0.100**	12.01 (3.58)	5.46 (2.54)	2.47 (1.86)	0.83 (1.33)	10.83 (3.38)	3.97 (2.17)	2.66 (1.87)	0.60 (1.28)	257.75
Carbosulfan 250 EC	**0.003**	12.33 (3.63)	4.01 (2.22)	1.61 (1.62)	0.52 (1.21)	10.83 (3.41)	3.44 (2.08)	1.60 (1.61)	0.29 (1.12)	263.80
Bifenthrin 10 EC	**0.002**	11.33 (3.46)	5.66 (2.55)	3.49 (2.08)	1.89 (1.67)	10.67 (3.39)	4.33 (2.30)	3.34 (2.05)	1.33 (1.52)	244.13
Neem oil	**0.050**	13.16 (3.76)	11.62 (3.53)	10.50 (3.39)	7.74 (2.97)	11.24 (3.48)	10.66 (3.41)	7.91 (2.51)	6.32 (2.79)	227.09
Control	-	14.00 (3.86)	13.17 (3.75)	12.04 (3.59)	11.04 (3.47)	11.66 (3.55)	11.19 (3.49)	10.09 (3.33)	9.66 (3.26)	215.78
CD (p dikhu 0.05)	-	NS	0.432	0.461	0.547	NS	0.484	0.666	0.395	
SE(m)		**0.151**	**0.144**	**0.154**	**0.182**	**0.196**	**0.162**	**0.223**	**0.132**	

*DBS: Days Before Spray; *DAS: Days After Spray.

Figures in parenthesis are square √x+0.5 transformed values.

11. Management of Wheat Aphid

Contact Scientist: Dr. Hafeez Ahmad

Methodology

Fields trials on wheat variety PBW-550 was conducted in the randomized block design to evaluate ten different insecticides including control for the management of wheat aphid. The experiments were carried out at Research Farm, Chatha, SKUAST-Jammu. The land preparation, seedling rate, fertilizer doses, irrigation and other cultural operations were performed as per package of practices of SKUAST-Jammu.

Table 1: Efficacy of different Insecticides against Wheat Aphid

Treatments	*Concentration*	*Average Yield (q/ha)*	*Increase in Yield Over Control*	*Value of Additional Yield (Rs)*	*Cost of Treatment (Rs)*	*Net Profit (Rs)*	*Cost Benefit Ratio*
Imidacloprid 200SL	0.009 per cent	32.65	11.40	25650.00	1150	24500.00	1:21.30
Thiamethoxam 25WG	0.045 per cent	31.72	10.47	23557.50	1980	21557.50	1:10.89
Dimethoate 30 EC	1.5 per cent	30.59	9.34	21015.00	815	20200.00	1:24.78
Acetamiprid 20 SP	0.1 per cent	24.87	3.68	8280.00	550	7730.00	1:14.05
Oxy demton methyl 25EC	1.5 per cent	28.48	7.23	16267.50	650	15617.50	1:24.02
Control	-	21.25	-	-	-	-	-

Recommendations

The technology developed proved effective in managing wheat aphid menace using imidacloprid followed by Thiamethoxam. The strategy of using imidacloprid gave satisfactory and reliable control of wheat aphids resulting in 34.02 per cent increase in yield having a Cost : Benefit ratio of 1:21.30 over control

12. Management of Leaf Folder in Paddy

Contact Scientist: Dr. Hafeez Ahmad

Methodology

Fields trials were conducted to evaluate ten different insecticides including control for the management of pod borer. The experiments were carried out at farmer's field. The land preparation, seedling rate, fertilizer doses, irrigation and other cultural operations were performed as per package of practices of SKUAST-Jammu.

Table 1: Economics of IPM against Leaf Folder

Treatments	*Concentration*	*Average Yield (q/ha)*	*Increase in Yield Over Control*	*Value of Additional Yield (Rs)*	*Cost of Treatment (Rs)*	*Net Profit (Rs)*	*Cost Benefit Ratio*
Clothianidin	0.06 per cent	62.75	24.48	14,688	1150	13,535	1:11.77
Betacyfluthrim + Chlorpyriphos	0.04 per cent	64.25	25.98	15,588	1750	14,838	1:8.48
Bacillus thuringiensis	500gm a.i/ha	54.65	16.38	9,828	1400	8,428	1:6.023
NSKE	5 per cent	49.25	10.98	6,588	500	6,088	1:12.18
Rope method	-	46.25	7.98	4,788	400	4,388	1:10.97
Rope method +Bt	-	56.75	18.48	11,088	1800	9,288	1:5.16
Rope method + NSKE	-	53.25	14.98	8,988	900	8,088	1:8.99
Control	-	38.27	-	-	-	-	-

Recommendations

The technology developed proved effective in managing leaf folder menace in rice. Betacyfluthrin + chlorpyriphos followed by Clothianidin were found to be best treatment against leaf folder control. Therefore, the Betacyfluthrin +

chlorpyriphos as conventional spray applications, followed by Clothianidin, are may be recommended for the control of leaf folder in paddy.

Benefits/Impacts

The insecticides used suppressed the pest population to economically non threatening levels.

Division of Plant Pathology

13. Round the Year Cultivation of Mushrooms

Contact Scientist: **Dr. Sachin Gupta**

Objectives

- Socio-economic upliftment of rural masses and unemployed youth
- Assured source of income through mushroom cultivation
- Utilization of infrastructure raised for mushroom cultivation

Methodology

Calendar for round the year cultivation of mushrooms using different types and species of mushrooms has been designed by the Division of Plant Pathology and the farmers/mushroom growers shall be advised to grow these mushrooms during the time period which is most conducive for the growth of that particular mushroom.

Table 1: Calendar of Year Round Mushroom Cultivation under Jammu Aubtropics

Type of Mushroom	*No. of Crops*	*Spawning*	*Termination of Crop*	*Cropping Period*
Agaricus bisporus (Button mushroom)	1st crop	Last week of September	Last week of December	8 weeks
	2nd crop	Last week of December	Last week of March	8 weeks

Type of Mushroom	*No. of Crops*	*Spawning*	*Termination of Crop*	*Cropping Period*
Pleurotus spp. (Dhingri)	1st crop	Second week of January	Last week of February	5 weeks
	2nd Crop	Last week of February	First week of April	5 weeks
Calocybe indica (Milky mushroom)	1st crop	First week of June	First week of August	8 weeks
	2nd crop	First week of August	Last week of September	8 weeks

Recommendation

Different mushrooms can be grown during the different time periods in Jammu division in such a manner that the resources meant for mushroom cultivation could be properly and effectively utilized and assured source of income generation from mushroom cultivation could be made.

Impact

Calendar for round the year cultivation of mushrooms have been adopted by many growers of Jammu region and this has assured their regular source of income throughout the year by means of mushroom cultivation.

Benefits

- Unemployed youth and rural masses can take up mushroom cultivation as an assured source of income
- Infrastructure raised by the mushroom growers could be effectively and properly utilized throughout the year for mushroom cultivation.

Division of Vegetable Science and Floriculture

14. Integrated Nutrient Management in Vegetable Crops

Contact Scientist: **Dr. Manoj Kumar**

Integrated nutrient management (INM) approach combines organic inputs with inorganic fertilizers to sustain soil fertility, improve plant productivity by considering total farm resources.

Methodology

Organic sources like Vermicompost, poultry manure and biofertilizers (PSB and Azotobacter) are used to supplement chemical fertilizers. They are use either as seed treatment, seedling dip or soil application. Doses can be adjusted as per the nutrient content availability.

Recommendations

Radish: Vermicompost @5.0T/ha + ½ N+P+K resulted in yield increase by 36 percent over control in cultivar Japanese White.

Cabbage: Vermicompost @2.5T/ha +75 per cent N+P+K + seedling dip with Azotobacter + PSB recorded yield increase by 40 percent over control in cultivar Golden Acre

Knol-khol: Azotobactor as seedling dip + 75 per cent N+P+K and full dose of FYM (15.0T/ha) resulted in yield increase by 40.0 percent in knol- khol cv. G-40.

Bottle gourd: Vermicompost @2.5T/ha + ½ N+P+K resulted in yield increase by 16.0 percent over control in cv. Punjab Komal.

Broccoli: Vermicompost @2.5T/ha + ½ N+P+K resulted in yield increase by 11.0 percent over control in cv. Early Green (Jammu Broccoli-07)

Table 1: Effect of INM Treatments on Yield of different Vegetable Crops

Sl.No.	*Crop*	*Recommended Treatment*	*Yield (q/ha)*		*Per cent Increase over Control*
			Recommended Treatment	*Control*	
1.	Radish	Vermicompost@ 5 T/ha + ½ N+P+K	123.2	90.0	36.89
2.	Cabbage	Vermicompost@ 2.5T/ha +75 per cent N+P+K + Seedling dip of *Azotobacter* + PSB	324.0	232.0	39.66
3.	Knol-khol	FYM @ 15T/ha + *Azotobactor*+75 per cent N+P+K	485.0	344.0	40.99
4.	Bottle gourd	Vermicompost @2.5T/ha+ ½ N+P+K	520.0	445.0	16.85
5.	Broccoli	Vermicompost @2.5T/ha+ ½ N+P+K	183.9	165.11	11.38

Control

Recommended dose of NPK.

Benefits/Impact

Integrated Nutrient Management improved soil health and reduced dose of chemical fertilizers by 25-50 per cent.

15. Portray Nursery Production Technology

Contact Scientist: Dr. Sanjeev Kumar

Plastic Portrays can be used for raising healthy nursery in high value low volume vegetables like cherry tomato, capsicum (coloured) and parthenocarpic cucumbers and in common vegetables (cole crops, exotic vegetables, brinjal, chilli, bitter gourd and bottle gourd) for early transplanting.

Methodology

Portrays are filled with potting mixture comprising Cocopeat, Vermiculite and Perlite (3:1:1). Alternatively vermicompost can also be used as low cost option by the farmer in place of standard potting mixture. A small depression is made in the middle of each cell and single seed is placed in each cell. It is then covered with a pinch of Perlite. The trays are kept in Polyhouse or in Verandhas. The growing media is moistened with water. Subsequently water soluble fertilizers in ratio of 19: 19:19 (N:P:K) are used as nutrient. In solanaceous vegetables and cucurbits the sowing is done in the 2nd week of December under protected conditions in cole crops during the end of August or early September.

Recommendations

Portrays (98 celled) can be used for solanaceous and cruciferous vegetables while 48 celled Portrays for cucumber and other cucurbits.

Benefits

The Portray Nursery Production Technology is suitable for raising healthy seedlings with 96-100 per cent germination success. The seedlings raised in Portrays result in early transplanting. The farmers can start harvesting the crop 20-25 days early and get better returns.

16. Seed Production of Knol-Khol under Non-Traditional Environment

Contact Scientist: Dr. Satis Kumar

Knol-khol is important crop and farmers are growing through periodical transplanting. The farmers require good quality seed of locally adapted varieties. With seed production technology developed at this institution the farmers can produce or save their own seed of locally adapted varieties.

Methodology

Knol-khol (G-40 or White Vienna) nursery is sown in August/September and transplanting is done in September/October at a spacing of 60 cm x 60 cm or 1 m x 1 m in flat beds in plains/low hills.

Seed multiplication can be done through *In-Situ* or Seed to Seed Method and *Ex-Situ* or Knob to Seed method. In the former method, plants are allowed to grow, over winter and produce seed in the original position where they were first transplanted. In the later method the mature plants after the formation of knobs are selected for true to typeness and knobs are uprooted. Leaves around the knob are removed while keeping the crown leaves intact. The selected knobs are replanted immediately in a well prepared field. For seed crop additional dose of 50 kg N, 50 kg P_2O_5 and 50 kg K_2O/ha. is given in case of *In-situ* and for *Ex-situ* the fertilizer are mixed with well rotten farm yard manure and filled in pits. The harvesting is done when 60-70 per cent of the crop turn brown. It gives an average seed yield of 639.17 kg/ha in *In-situ* and 445.25 kg/ha in case of *Ex-situ* method of seed production.

Recommendation

For nucleus and breeder seed production *Ex-situ* method of seed production is recommended. For foundation, certified and farmer saved seed *In-Situ* method is recommended in plains/low hills for above specified variety.

Benefit

Quality seed can be produced by the line departments and farmers following *In-situ* method, since isolation requirement problems can be taken care in non traditional environment.

17. Rapid Multiplication of Tomato Hybrids through Plant Tissue Culture

Contact Scientist: Dr. Anil Bhusan

Hybrids have become very important in Tomato all over the world. Hybrid seed production is possible through traditional methods provided parental lines are available. Protocol has been standardized for large scale multiplication of public sector tomato hybrids that can have application in newly released/developed hybrids

Methodology

Adventitious Shoot Initiation/Emergence

The cotyledons as explants when cultured on MS medium using different concentrations of BAP, maximum shoot induction/emergence response after 4 weeks was observed on medium supplemented with 2.0 mgl^{-1} BAP.

Shoot Multiplication

Best response towards shoot multiplication was observed on MS medium fortified with 2 mgl^{-1} BAP. The maximum number of shoots per explant was observed in public sector hybrids TH802 (5.17) and TH2312 (5.11).

Rooting

Maximum number of roots per shoot and root length were recorded in ½ MS medium without hormones or low level of auxins in both hybrids

Hardening

Micropropagated plantlets were hardened by continues exposure to reducing humidity and increasing light intensity for better survival. The treatment of plantlets with *viz.* Glycerol @0.5 per cent or Abscisic acid @ 5ppm before hardening improved survival as compared to control in acclimatized plants. Plantlets after hardening in the pots containing Sand, Soil and Compost (1:1:1) in a green house were transferred to the field.

Benefit

By using above reproducible protocol, rapid and large scale multiplication of multi disease resistant institutional or public sector is possible since seeds are very expensive. Plant tissue culture can also be taken up by unemployed educated youth or farmers for self-employment purpose

Shoot Initiation/Emergence and Shoot Multiplication in Tomato.

Rooting on Basal Medium in different Genotypes in Tomato.

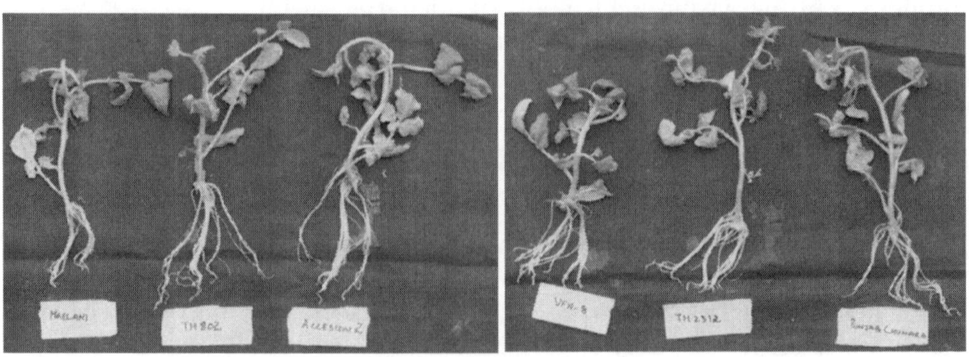

Rooted Plantlets of Tomato Genotypes.

18. Rapid Multiplication of Ginger Planting Material by Low Cost Protocol (PTC)

Contact Scientist: Dr. R.K. Gupta

Methodology

Low cost effective micro propagation protocol for ginger has been standardized using ordinary Isubgol as a gelling agent in place of expensive Agar agar table sugar in place of sucrose and tap water in place of distilled water using MS medium. For initiation of initial aseptic cultures(I^{st} step) MS-medium supplemented with cytokinin 5-10 mg^{-1} + sucrose 3 per cent is used. Shoot multiplication (2nd step) can be done MS-medium solidified with Isubgol and supplemented with BAP 2.0 $mg1^{-}$ + NAA 0.5 mg^{-1} + using Ordinary sugar (3 per cent). Further multiplication and maintenance of cultures in various cycles was possible on basal MS medium with lower doses of BAP (0.5- 1.0 mg^{-1}) or even on hormone free medium. The cultures can be maintained in indirect natural light near Windows or diffused light under shade of trees while maintaining aseptic conditions. The shoot and root formation occurs simultaneously. There is no need of *ex-vitro* rooting. The micropropagated plantlets derived from Isubgol gelled medium perform relatively better over the ones derived from Sago or agar gelled medium. Performance of rhizomes (2nd generation) and (3rd generation) was also recorded as better when compared with crop raised from conventional derived rhizomes.

Recommendation

The low cost Micropropagation protocol standardised is reproducible and cost effective and can be used for multiplication of locally adapted varieties or elite line which otherwise will take over 10 years for multiplication of new varieties or lines since ratio in conventional method is 1: 4 or 5.

Benefit

Rapid and large scale multiplication of disease free locally adapted ginger or rejuvenation of deteriorated indigenous material can be taken up for revival of varieties by educated youth or young educated farmers or field functionaries.

19. Rapid Potato Multiplication through Micro-Tubers Methodology

Contact Scientist: **Dr. R.K. Gupta**

Methodology

Performance of potato seed tubers (micro tubers) over conventional tubers was compared in potato genotypes (Kufri Badshah and Kufri Pukhraj). Micro tubers of tested cultivars performed well in the open field and performance was variety dependent. With regard to responses to potassium application to n mini or normal seed sized tuber, data revealed that maximum mean average tuber weight (140.06 g), equatorial diameter (100.37 mm) was observed under higher dose of potassium (150 kg/ha) and these values were comparable with the next lower dose of potassium (120 kg/ha) with exception of polar diameter.

Recommendation

A new method is advocated for PTC mediated disease free potato seed multiplication that can have commercial applications for spread of new varieties. Progressive seed potato growers can use plant tissue culture derived micro or mini tubers for further multiplication in open or in insect proof net house.

Benefit

Rapid and large scale multiplication of disease free potato micro/mini tubers can be taken up by farmers as Potato seed tuber production venture especially for new varieties, hybrids and local germplasm.

20. River Bed Production of Early Vegetables

Contact Scientist: **Dr. R.K. Samnotra**

Introduction

River bed production of early vegetables in trenches is a very old practice of growing vegetables on the basins of the rivers where sub surface water is being utilized by the crops during its life cycle. The capillary rise is known to be major source of meeting the water requirement of crops under shallow water table conditions. Refinement in trench orientations, trench geometry, angles of wind breakers has extended vegetable production on larger areas otherwise lying barren on the river banks.

Objectives

- To utilize underground water for production of early/off season vegetables on the banks of perennial rivers
- Efficient use of barren lands on the river banks
- To develop trench techniques on scientific lines
- To identify vegetables suitable for out of season production under trenches

Methodology

Development of Trench Techniques for Production of Early Vegetables on the River banks

1) Trench Orientations
2) Trench geometry at sowing
3) Orientation of wind breakers

Trench Orientations:

T_1) North-East, T_2) North-West and T_3) East-West

Trench Orientations

Results

The North-East orientation was found to be the best in all the cucurbits in which less mortality of the seedlings during the intense cold conditions was recordedresulting in to a greater number of female flowers and higher yield than the other two directions under study.

Trench geometry

T_1) 25m x1.0m x Ground water availability level

T_2) 30m x 1.0m x Ground water availability level

T_3) 35m x1.0m x Ground water availability level

Trench Geometry (LxWXH)	*Cucumber (Hybrid) Yield q/ha*	*Bitter gourd (Hybrid) Yield q/ha*	*Bottle gourd Hybrid (Long) Yield q/ha*	*Indian Squash Yield q/ ha*	*Summer Squash Yield q/ ha*	*Bottle gourd (Round) Yield q/ ha*
25mtx1.0mtx WA	170.60	560.25	800.40	210.50	700.10	680.00
30mtx 1.0mtx WA	195.25	750.50	990.33	265.25	860.25	823.56
35mtx 1.0mtx WA	165.70	590.60	850.50	240.40	630.50	810.25
CD 5%	26.21	35.28	45.56	12.27	32.25	47.04
Net Returns(Rs.)	3.80,000	3.74,000	4,44000	3,00000	3,20000	3,90000

Results: Trench geometry of 35m x1.0m up to ground water availability level was found to be the best which out yielded significantly in all the cucurbits than the other two trench geometries under study.

Orientation of wind breakers

To protect the crop in its early stages from intense cold waves and frost conditions from December to January months the orientation of wind breakers is of utmost importance for successful production of early crops.

Angle Orientations of windbreakers: $60^{0,} 45^{0}$ and 30^{0}

Windbreaker angles	Cucumber (Hybrid) Yield q/ha	Bitter gourd (Hybrid) Yield q/ha	Bottle gourd Hybrid (Long) Yield q/ha	Indian Squash Yield q/ ha	Summer Squash Yield q/ ha	Bottle gourd (Round) Yield q/ha
60^0	190.33	800.00	770.33	275.50	750.35	768.33
45^0	168.52	750.50	962.25	212.35	655.33	742.25
30^0	145.20	670.32	710.10	180.33	570.25	590.33
Mean	167.35	740.27	814.22	668.18	658.64	700.30
CD 5%	21.03	60.66	57.09	18.55	95.09	33.07
Net Returns	3.80,000	3.74,000	4,44000	3,00000	3,20000	3.90,000

Results: Wind breakers at an angel of 600 were found to be the best to protect the seedlings from frost and cold waves during intense cold conditions. Minimum mortality of the seedlings was recorded at this orientation as compared to other two angels under study. This angel also facilitates inter cultural operations in a better way and was found to be at par with the treatment comprising of two-tier protection from frost.

Impact Outcome

- ❒ Increase in area under vegetable production
- ❒ Availability of early/off season vegetables in the market
- ❒ Rise in socio-economic status of the farmers

21. Knol-Khol Production Round the Year

Contact Scientist: Dr. R.K. Samnotra

Knol-khol (*ver.*Karam/Ganth gobhi) is scientifically *Brassica oleracea* L.var. *gongyloides* belongs to family Brassicaceae. It is a cool season vegetable crop grown in rabi season (October-March) under subtropical conditions for both its knob and leaves. But due to scientific intervention and varieties, it can be grown during summer and rainy months for its leaves which remain highly in demand among the consumers.

Objectives

1. Round the year production of knolkhol
2. Increase in cropping intensity (500 per cent)

Methodology

A. Transplanted Crop

Knol-khol is normally cultivated by nursery raising followed by transplanting and four crops are obtained through this method. However, during rainy season, 5th crop which is sown directly in the field without nursery preparation can be taken. This crop is taken for leaves and small swollen stems, which are highly preferred by the consumers and fetches high price in the market. The crop took 30-40 days for harvesting.

The time of nursery sowing, transplanting and direct sowing of knolkhol in the field under sub-tropical zone of Jammu is given under:

Zone- Sub-tropical	*Sowing Time*	*Transplanting*	*Varieties*
1st crop	Mid September	Mid October	G-40
2nd crop	Mid October	1st week December	White Vienna
3rd crop	Mid February	Last week March	Purple Vienna
4th crop	Mid April	Mid May	King of Market
5th crop	**Last Week August**	**Direct sowing**	

Direct Sowing

Direct sowing of crop in lines 30 cm apart by kera method can also be done by the end of rains (Last week of August). Proper soil moisture is must for direct sowing. Thin the crop by keeping 20 cm distance between the plants after one month. The plants so obtained by thinning can further be transplanted in other prepared fields in the month of October.

Seed rate	**Transplanted crop**	1-1.5 kg/ha
	Direct sowing	4-5kg/ha.
Fertilizer	**Transplanted crop**	20 tons of well rotten FYM or compost + 100 kg N, 50 kg P_2O_5 and 50 kg K_2O/ha
	Direct sowing	20 tons of well rotten FYM or compost +60 kg N, 30 kg P_2O_5 and 30 kg K_2O/ha
Irrigation	**Transplanted crop**	Light irrigation should be given just after transplanting of seedlings and subsequent irrigations should be given at weekly interval during winters and 3-4 days during summers.
	Direct sowing	In direct sown crop, proper soil moisture is required for effective germination. Proper drainage must be provided to the crop during rainy season
Hoeing/ Weeding	**Transplanted Crop**	First hoeing should be done after 10 days of transplanting and second hoing and weeding should be done after 20-25 days after transplanting
	Direct sowing	In direct sown crop, thinning of seedlings should be done after 15 days after sowing to maintain proper plant population followed by weeding and hoeing after 10-15 days after thinning
Harvesting	**Transplanted crop**	It should be harvested when the knobs are tender and have attained 5-7 cm diameter. This crop took 40-45 days after transplanting.
	Direct sowing	The direct sown crop is harvested when the plants attained full vegetative growth with small swollen stems. This crop does not show proper knob formation and took 30-40 days.
Yield	**Transplanted crop**	200-250q/ha
	Direct sowing	100-150q/ha
Plant protection	**Transplanted crop**	Seeds should be treated either with captan or thiram @3g/kg before sowing in the nursery. To manage cabbage butter-fly and cabbage caterpillar, spray the crop with Thiodon 35EC @1.5 l in750 litres of water. If attack of aphids or cabbage saw fly is observed spray the crop with Melathion 50 EC @ 1.5 Litres in 750 litres of water.
	Direct sowing	Seeds should be treated either with captan or thiram @3g/kg before sowing in the field. Soil should be treated with chlorodane dust or Lindane dust @ 35-50 kg/ha to avoid the attack of cut-worms and white grubs.

B. Transplanted Crop

Knol-khol is normally cultivated by nursery raising followed by transplanting and four crops are obtained through this method. However, during rainy season, 5th crop which is sown directly in the field without nursery preparation can be taken. This crop is taken for leaves and small swollen stems, which are highly preferred by the consumers and fetches high price in the market. The crop took 30-40 days for harvesting.

The time of nursery sowing, transplanting and direct sowing of knolkhol in the field under sub-tropical zone of Jammu is given under:

Zone-Sub-tropical	*Sowing Time*	*Transplanting*	*Varieties*	*Seed Rate (kg/ha)*
1st crop	Mid September	Mid October	G-40	1-1.5
2nd crop	Mid October	1st week December	White Vienna	1-1.5
3rd crop	Mid February	Last week March	Purple Vienna	1-1.5
4th crop	Mid April	Mid May	King of Market	1-1.5
5th crop	**Last Week August**	**Direct sowing**	All varieties	4-5.0

C. Direct Sown Crop

In the end of summer season and before the onset of rains, the field is well prepared with a gentle slope. In the last week of August, seed is broadcasted in the field having optimum moisture condition. Thinning of over-crowded seedlings after 21 days of sowing along with hoeing and weeding is done. Keeping plant to plant distance upto 15 cm. Uprooted seedlings are also transplanted in field as a relay crop of knolkhol to meet the local market demand.

Impact Outcome

1. Production of offseason knolkhol crop fetches higher returns to the farmers
2. Increase in farm employment due to higher cropping intensity and efficient land use
3. Availability of the crop to the consumers round the year.
4. Rise in socio-economic status of the farmers

Division of Sericulture

22. Chawki Rearing Technology for Increasing Cocoon Production

Contact Scientist: **Dr. R.K. Bali**

Objectives

- Development of healthy and uniform young age larvae.
- Development of robustness in newly hatched silkworm larvae.
- Improvement in quality and quantity of cocoon production.

Introduction

Rearing of larvae upto second age is termed as chawki. The main aim of conducting chawki is to have uniform growth of worms for higher cocoon yield. The rearing of young age larvae is conducted at chawki rearing centers organized under expert technical supervision.

Methodology

- For rearing 100 dfls of chawki worms, 120 sq. ft. rearing space and ten plastic trays of 3'x 2'x 3' required.
- Half an acre of mulberry garden under irrigation is sufficient for brushing and rearing of 100 dfls of eggs.

- Chawki worms being delicate are susceptible to bacterial and viral diseases and requires utmost hygienic conditions during rearing.
- The newly hatched worms should not be starved for undue long period.
- Top 2-3 full blown leaves immediately below the growing bud are recommended for chawki rearing.
- Mulberry leaves must be plucked during cooler hours and transported to rearing house in wet gunny cloth/basket/plastic crate. For storage of leaves cooler place in the rearing house is chosen and humidity maintained close to nearly cent per cent.
- Newly hatched larvae should be fed with finely chopped (0.5-lcm^2) tender and succulent mulberry leaves of variety Chinese white/S146/Victory -1/S-1635.
- Keep worms and leaf in a thin bed surrounded by wet foam pads.
- Cover rearing trays with paraffin or newspaper for moisture/humidity retention.
- For chawki rearing approximately the size of the leaf and environmental requirement for uniform growth is as under:

Stage	*Leaf Size (cm^2)*	*Temp (°C.)*	*Humidity*	*Number of Feeds*	*Number of Cleaning/Stage*
I	0.5-1.01.5-2.0	28	90	4	1
II	2.0-3.03.0-4.0	28	90	4	1

- The optimum spacing for young age worms in first two instars for 100 dfls is as under:

Age of Worms	*Area of Rearing Bed*		*Increase in Space*
	At Beginning	*At End*	
I	4 sq ft	14 sq ft	3 ½ times
II	15 sq ft	45 sq ft	3 times

- During rainy season, the atmospheric humidity being high, wet paper/ foam pads and piling of trays should be avoided.
- Prior to every feeding, the paraffin/newspaper and foam pads should be removed to allow the rearing bed to dry up quickly.
- The leaf requirement for 100dfls of bi x bi hybrid during first age is around 2.5 to 3.0 kg and 8.0 to 9.0 kg for second age.
- In first age, one bed cleaning prior to setting of worms for first moult and

twice in second instar, one at the time of resumption for second instar and second around 12 hours prior to settling for second moult.

- During settling of worms for moult, the top paraffin/newspaper and foam pads should be removed.
- During moulting, worms should not be disturbed.
- Chawki reared worms should be transported to rearers place during cooler hours.

Table 1: Year/Season-wise Cocoon Crop Production

Year	*Seed Reared (Oz.)*			*Average Cocoon Yield/Oz.*		*Average Cocoon Yield/ Rearer (Kg)*
	Spring	*Autumn*	*Total*	*Spring*	*Autumn*	
I (2013)	43.00	8.5	51.5	47.000	30.500	38.750
II (2014)	48.25	21.0	69.25	55.230	37.459	46.389
III (2015)	70.50	15.5	86.0	57.872	40.829	49.350
3 years	161.75	45.0	206.75	53.397	36.262	44.829

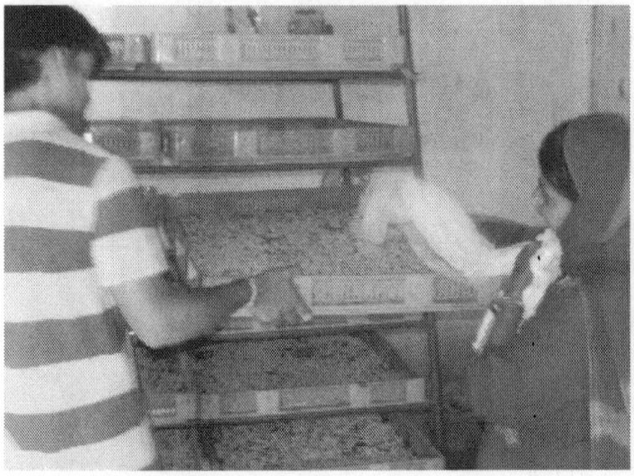

Recommendations

Chawki reared worms (after 2nd moult) should always be distributed among silkworm rearers particularly in harsh climatic conditions (autumn season).

Benefits/Impact

Distribution of Chawki reared worms showed substantial improvement in cocoon productivity and quality. Second commercial additional rearing in autumn season resulted into increase in total cocoon production with substantial increase in the income of adopted silkworm farmers. On the basis of three years averages, cocoon production of 53.397 kg/Oz. of silkworm seed in spring rearing season and 36.262 kg/Oz. of silkworm seed in autumn rearing season was achieved. Distribution of chawki reared worms showed a visible impact of 42.3 per cent increase on cocoon crop productivity during spring and 31.8 per cent in autumn season.

Season	*Average Yield of Distt. Udhampur (Kg/oz) for Non-chawki Reared Worms*	*Average Cocoon Yield (Kg/Oz) for Chawki Reared Worms*	*Average Increase in Cocoon Yield (Kg/oz) over Non-chawki Reared Worms*	*Benefit (Rs.)/ Rearing @ Rs.500*
Spring	37.5	53.521	16.021	8000
Autumn	27.5	35.639	8.139	4000

23. PO3 x ND5: A Productive Bivoltine Hybrid

Contact Scientist: Dr. R.K. Bali

Objective

- Development of season and region specific bivoltine hybrid.

Introduction

The Jammu division of the J&K state falls under western Himalayas constituting,

- Outer plains (major parts of Kathua and Jammu districts),
- Outer hills (lower reaches of Udhampur Poonch and Rajouri and upper reaches of Jammu and Kathua districts)
- Middle mountains (mid and high altitude of Poonch, Rajouri, Kishtwar and Doda districts).

The pragmatic agro climatic diversification that exists across the physical divisions of the state, delimits the state in distinct sericultural zones. On the basis of sericultural cropping schedules, Jammu division possesses two distinct sericultural zones temperate and sub –tropical areas. Whole of Doda, Kistwar, Poonch and higher reaches of Udhampur and Rajouri districts of Jammu fall under temperate zone while as district Jammu, Kathua, Samba, Reasi, Ramban and foot hills of Udhampur and Rajouri fall under sub tropical sericulture belt. The mulberry plantation is mostly available in the form of roadside trees. The concept of raising nurseries in the private sector is lacking The silkworm rearing seasons in temperate zone are May/June (Spring) while as for subtropical zone rearing is conducted in March/ April. The authorized silkworm combinations are commercially used yielding around 30-35kg/oz of seed during spring and 15-20 kg/oz of seed during autumn.

Methodology

The exotic Japanese hybrids were used as donor parental genetic material for the development of inbred lines. Ten commercial economic traits were fixed during twelve generations. The eggs of the hybrid are *diapausing* in nature. Single mother moth lays 450 -550eggs. The silkworm larvae of the hybrid are marked. The cocoons are slightly constricted in shape, compact in feel with smooth wrinkles. Cocoon and yarn colour of parental inbred lines and hybrid is white.

Table 1

Parameters	*Existing Hybrid in Field SH6 x NB4D2*	*New Hybrid PO3 x ND5*
Survival per cent	90	90
Cocoon Yield/100 Dfls (per kg)	55.00	70.00
Cocoon weight (g)	1.75	1.90
Cocoon shell weight (g)	0.33	0.41
Cocoon shell ratio (per cent)	18.85	21.58
Filament length (m)	610	970
Grade of Silk	Ungraded	A- Grade

Race Authorization

Hybrids PO3 x ND5 evolved at Division of Sericulture, SKUAST Jammu got entry for evaluation at National level during 2011-12. Seven hybrids from APSSRDI, Hindupur (AP), CSR and TI (Mysore), KSSRDI (Banglore) and SKUAST-J (Jammu) were evaluated. The pooled data generated by the North Zone Race Authorization Test Centers is presented as under shown in Tables 2 and 3.

Recommendations

Based on seasonal and regional performance in the northern states by the Race Authorization Committee of Central Silk Board hybrid PO3 x ND5 can be recommended for temperate regions of Jammu Sericultural zones.

Table 2: Performance of Bivoltine Hybrids during Spring

Hybrid	*Fecundity*	*Yield 10000 Larvae No.*	*Yield 10000 Larvae wt./kg*	*Single Cocoon wt./g*	*Single Shell wt./cg*	*Shell Ratio per cent*	*Filament Length m*	*Denierd*	*Raw Silk per cent*	*Reel-ability per cent*
(AP71xAP9) x(AP72xAP8)	655	9568	14.474	1.531	31.03	20.35	846	2.498	35.54	82.90
(CSR50)x(CSR52) X(CSR51)x(CSR53)	615	9661	15.960	1.588	36.99	23.26	952	2.759	39.34	84.30
(CSR6 x CSR26) x (CSR2 x CSR27)	596	9796	16.218	1.581	35.77	22.65	887	2.719	39.69	80.90
AP71xAP72	534	9605	14.918	1.461	31.14	21.31	892	2.507	35.20	80.70
CSR50 x CSR51	520	9542	16.128	1.559	36.70	23.77	1006	2.814	39.90	83.00
CSR21DR x CSR28DR	429	9727	15.024	1.463	32.31	21.97	915	2.800	38.69	83.50
PO3 x ND5	451	9758	17.018	1.725	40.37	23.39	813	2.835	36.64	79.50
NK2 x HND	582	9543	15.946	1.508	31.32	20.70	810	2.510	34.86	80.30
CSR2 x CSR4	507	9769	16.122	1.522	33.05	21.70	875	2.721	36.05	82.30

Table 3: Performance of Bivoltine Hybrids during Autumn

Hybrid	*Fecundity*	*Yield 10000 Larvae No.*	*Yield 10000 Larvae wt./kg*	*Single Cocoon wt./g*	*Single Shell wt./cg*	*Shell Ratio per cent*	*Filament Length m*	*Denierd*	*Raw Silk per cent*	*Reel-ability per cent*
(AP71xAP9) x(AP72xAP8)	585	8940	15.600	1.610	34.23	21.27	987	2.646	36.25	82.00
(CSR50)x(CSR52) X(CSR51)x(CSR53)	542	9542	17.334	1.746	40.18	22.99	1082	2.761	39.24	79.50
(CSR6 x CSR26) x (CSR2 x CSR27)	511	9755	18.168	1.755	40.20	22.85	957	2.775	40.24	83.90
AP71xAP72	558	8785	14.852	1.585	34.00	21.45	924	2.549	34.61	80.30
CSR50 xCSR51	514	9590	17.626	1.725	40.51	23.54	1028	2.658	40.30	84.50
CSR21DR x CSR28DR	448	9533	17.390	1.572	30.78	19.59	956	2.641	35.69	78.40
PO3 x ND5	495	9687	17.240	1.875	43.74	23.34	1041	2.660	37.38	80.30
NK2 x HND	474	9696	18.468	1.838	38.75	21.10	890	2.749	33.93	77.40
CSR2 x CSR4	541	9786	16.572	1.746	42.68	24.42	910	2.806	36.92	80.20

Impact/Benefit

The hybrid was validated through KVK Kathua at field level and on an average 5 kg cocoons per 100 dfls extra yield was harvested in comparison to ruling hybrid given by Sericulture Development Department and each farmer got a benefit of Rs.3000 extra per rearing.

Division of Soil Science and Agricultural Chemistry

24. Soil Fertility Mapping of Jammu district

***Contact Scientist:* Dr. Vikas Sharma**

Objectives

- To prepare database of soil available nutrients in relation to soil properties
- To generate soil fertility maps through GIS technique

Introduction

GIS (Geographical information system) technique is now widely used to delineate variation in soil properties as well as their mapping, due to its advantages over earlier methods. Soil testing provides information regarding nutrient availability in soils which forms the basis for the fertilizer recommendations for maximizing crop yields. Information on soil properties has often been sought for agronomic applications and was obtained from random sampling within defined areas such as plots, fields, and farms. However, with the help of modern equipment came the ability for strategic sampling and subsequent spatial representations of various soil properties. A soil fertility map for a particular area can prove highly beneficial in guiding the farmers, manufacturers and planners.

Research Methodology

The Jammu district of Jammu and Kashmir covers an area of 2336 sq. km. and falls in the sub-tropical zone of the state. Soil samples were collected from the entire Jammu district following the stratified random manner so as to ensure suitable representation. Each sample was geo-tagged with the help of a GPS (Global Position System). Soil samples were processed for analysis of basic soil properties and nutrient status as per standard procedures. The process of digitization and generation of maps was carried out with ArcGIS 9.2 software following ordinary kriging.

Results

Wide variations in soil properties have been observed. Topography, vegetation and also management practices seem to be contributing to this variability. The GIS soil fertility maps provide clear information on the patterns and trends with respect to individual soil properties. The pH of soils in the Jammu district varied from acidic to alkaline. The southern plains exhibiting alkaline conditions and the northern foothill soils leaning towards slightly acidic conditions. pH below 6.5 has been observed in hill soils which were largely under pine vegetation. As far as electrical conductivity (EC) is concerned, no disturbing trends were observed, as most of the soils had EC values of less than 0.4 dS/m. These values are much below the threshold value of 4 dS/m, beyond which the soils are classified as saline. Interesting variations with respect to organic carbon (OC) content were observed. Based on the limits for OC *i.e.* low (<0.50 per cent), medium (0.50-0.75 per cent) and high (> 0.75 per cent), the soils were in general low in organic carbon content in the cultivated *kandi* areas of Vijaypur, Purmanadal, Dansal and Khore blocks. The northern areas of the Jammu district had relatively higher organic carbon content owing to the natural agro-forestry systems as well forests here. In fact more than 1.0 per cent organic carbon was observed in soils under the forests. In the southern blocks of RS Pura and Bishnah, it varied between medium to high. On the whole, majority of the soils of the district were low (<280 kg/ha) in available nitrogen. Areas showing medium (> 280 kg/ha) N content could be largely because of management practices (relatively high application of N fertilizers). The areas under low N content were subdivided further to get a clear picture of the N status. As in case of OC, the cultivated and degraded *kandi* areas had N content of less than 190 kg/ha. Available P content was largely medium with some areas in Khore block showing low content. Available potassium levels were largely in the medium category. As far as micronutrients are concerned, the entire district was deficient in DTPA extractable Zinc (Zn), with values falling well below the critical limit of 0.6 ppm for soils. Deficiency of other micronutrients was not observed in the Jammu district, although wide variation in their content was observed. DTPA extractable manganese showed strong relation with pH, with low pH soils in northern areas of Jammu district showing Mn content of > 15 ppm and in some cases exceeding upto 45 ppm. More than 60 per cent of area had less than 1.00 ppm copper, however, only negligible samples were below the critical limit of 0.2 ppm.

Soil organic carbon map of Jammu district

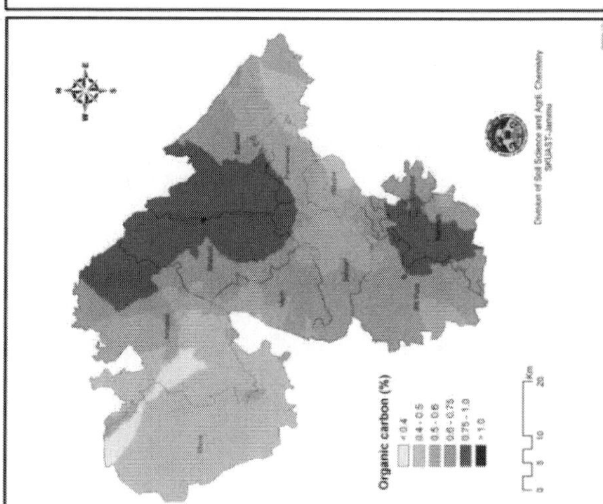

Soil pH map of Jammu district

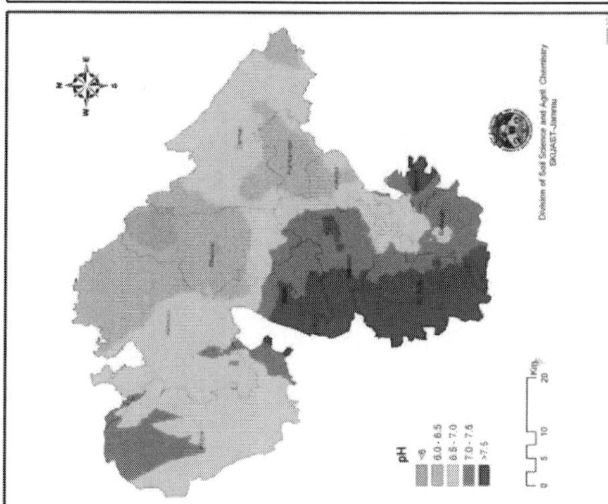

Electrical conductivity of soils of Jammu district

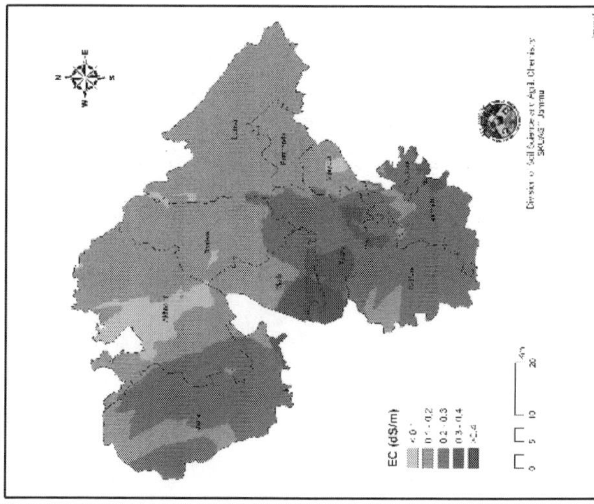

Recommendations

A soil fertility map for a particular area can prove highly beneficial in guiding the farmers, manufacturers and planners. The Northern cultivated areas of the district were poor in organic carbon and nitrogen, indicating that application of organic matter and balanced use of nitrogenous fertilizers needs to be practiced. The zinc deficient blocks of the south need applications of $ZnSO_4$ for optimum rice productivity.

The maps have been supplied to KVKs and line departments.

25. Assessment of Physical Health and Fertility Status in Soils of Udhampur District of Jammu

Contact Scientist: **Dr. Piyush Sharma**

Objectives

1. To assess the physical health and nutrient status of soils of Udhampur district.
2. To generate digital maps of soil physico- chemical properties using Geographic Information System (GIS).

Introduction

Soil can be defined as a highly complex assemblage of geological materials, dead organic matter, living roots, soil animals and microbes, soil water and soil air. Together they provide the physical and chemical conditions necessary for plant life, and consequently for most forms of animal and microbial life. GIS based fertility management is an effective tool for increasing productivity of agricultural soils that have a high degree of spatial variability.

Methodology

Study area lies in between 32.93°N of latitude and 75° E of longitude. The climatic of Udhampur is warm and temperate having average temperature 21°C. Annual rainfall is 1463 mm. The summer temperature as high as 38 degree Celsius in the plains and winter temperature in the upper hilly areas touches sub-zero. For the purpose of this study a total of 248 surface soil samples (0-15cm), representing agriculture, horticulture, forest, pasture, and wasteland were collected randomly from different elevation across whole of the district by using global positioning systems (GPS). Maps for the study area were prepared using geo processing tools of ArcGIS 10.3 software. Collected soil samples were air-dried and grinded with the help of wooden hammer.

Maps

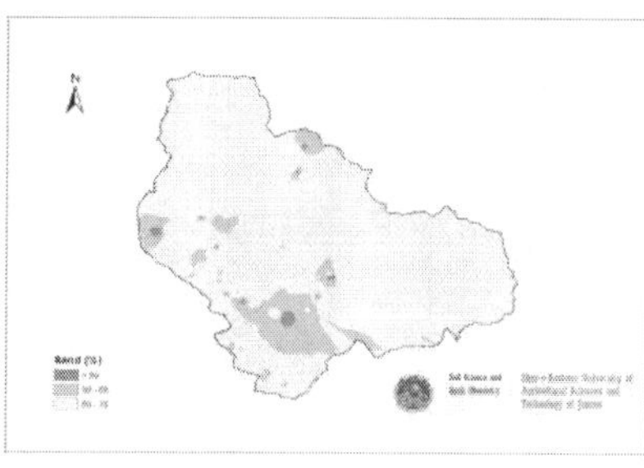

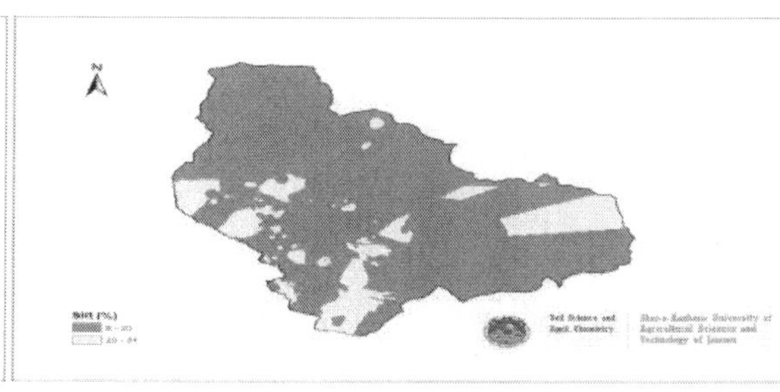

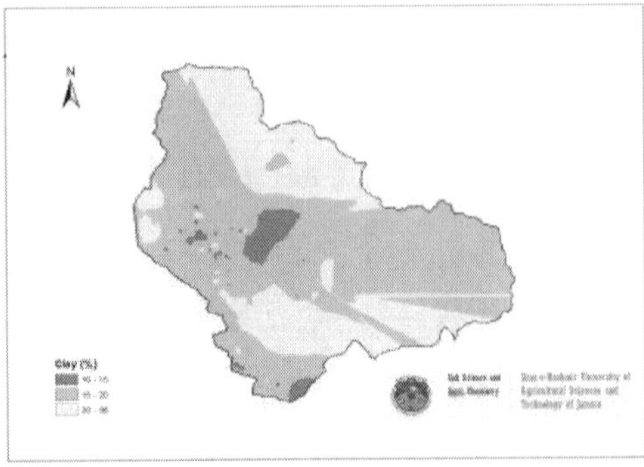
N
Clay (%)

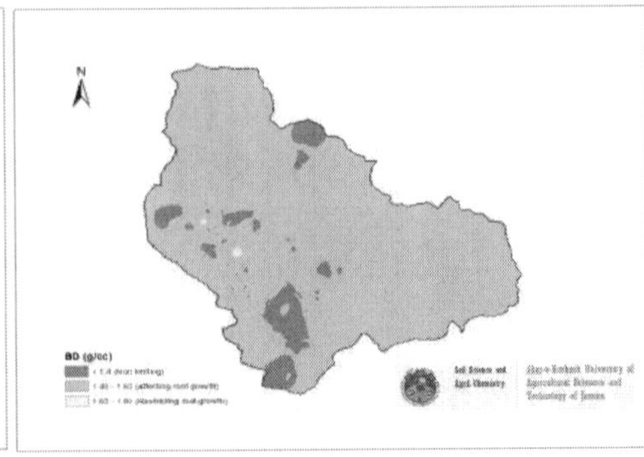
N
BD (g/cc)

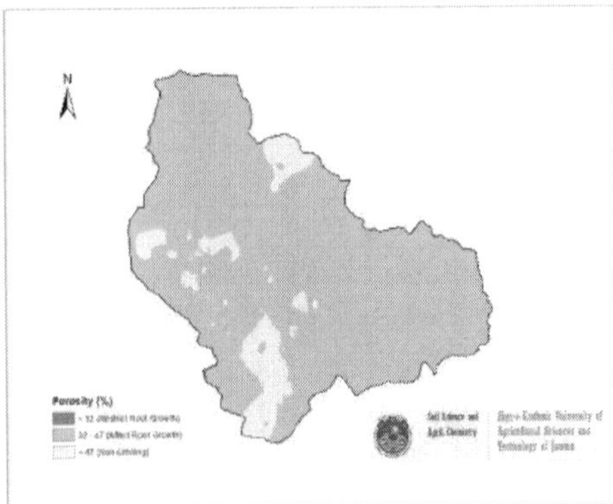
N
Porosity (%)

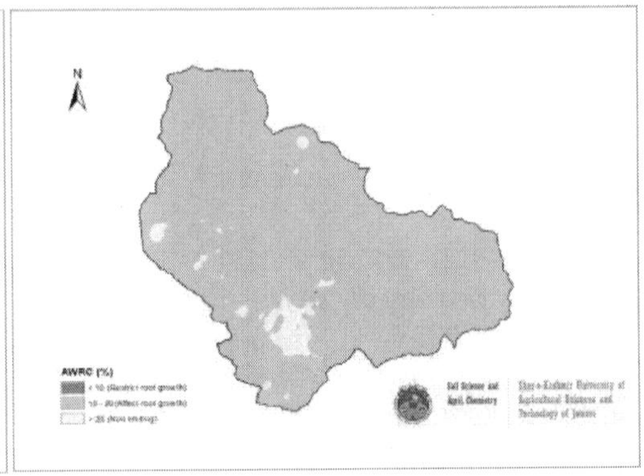
N
AWRC (%)

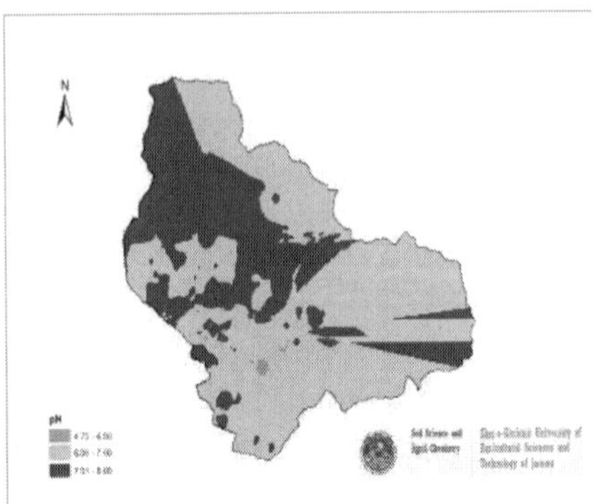
N
pH

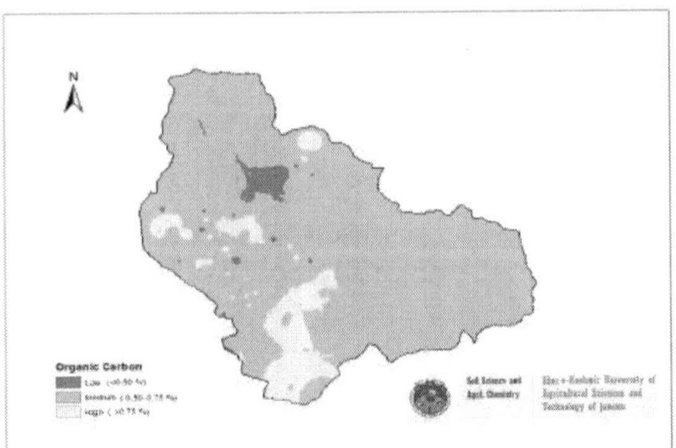
N
Organic Carbon

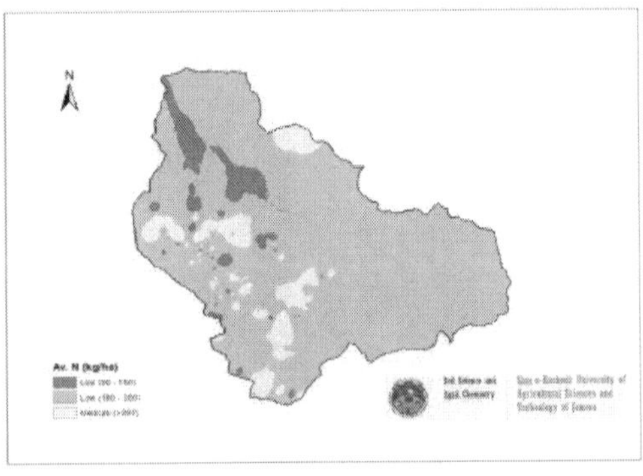
N
Av. N (kg/ha)

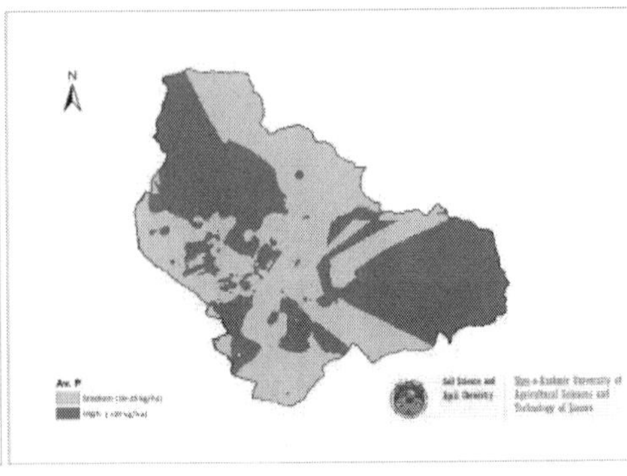
N
Av. P

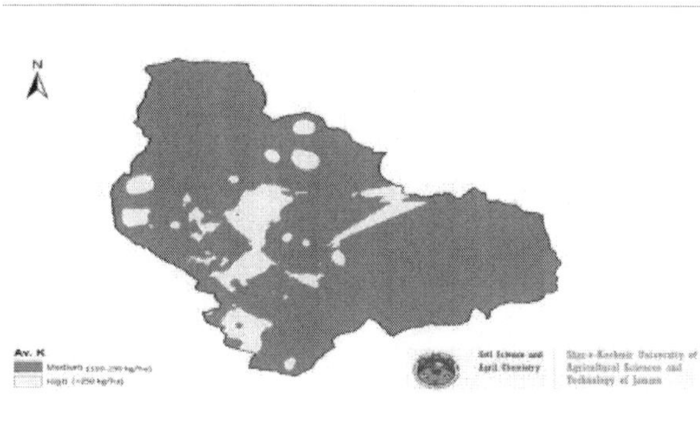

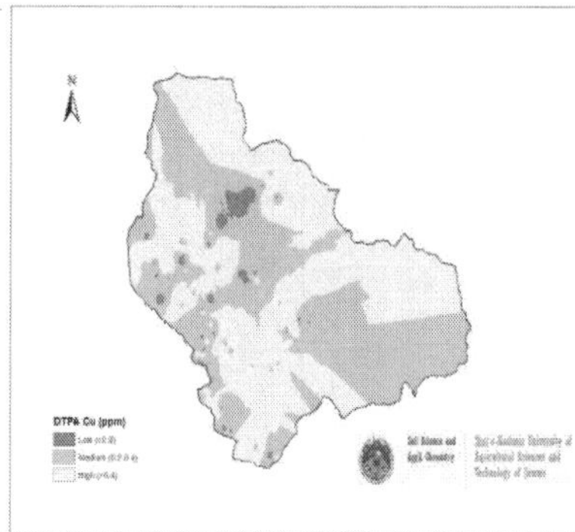

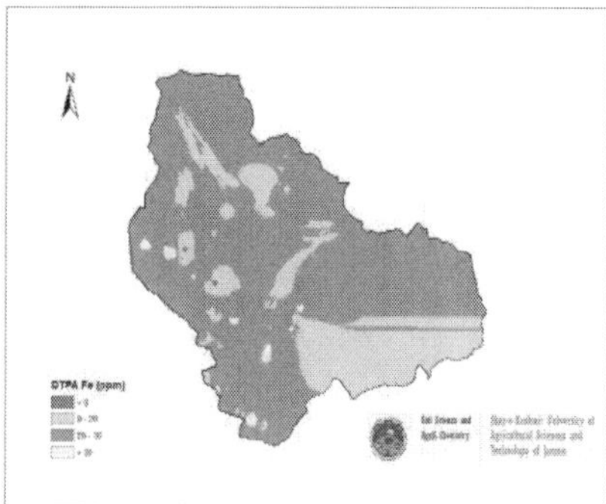

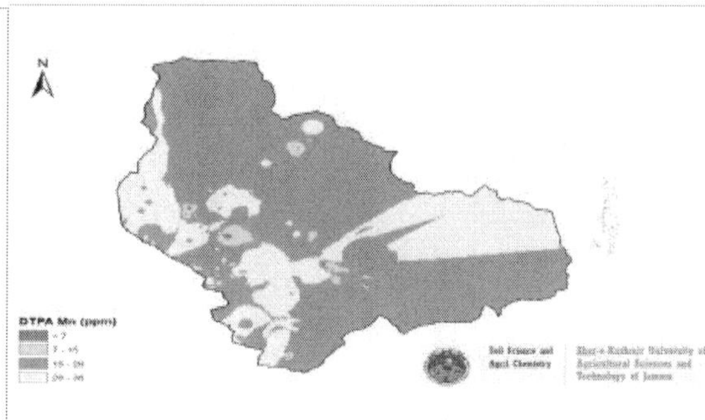

Findings

- GIS based soil fertility maps of Udhampur district were developed. The predominant textural class was sandy loam. Soil pH of major parts of the district ranged from 6.0 to 7.0. In most of the areas of Udhampur district. Soil organic carbon was in the medium range except some parts of Majalta and Ramnagar blocks. Available N is low, available P ranged from medium to high and available K is mostly in the medium range. Copper, iron and manganese were sufficient in the soils of entire district. Whereas soils of about 80 per cent portion of the district were found to be deficient in zinc content.
- Soil physical properties under different land use systems indicated that physical health was relatively better in forest and pasture soils as compare to the other land use system.
- Under different land use system of Udhampur district, forestland was found to be rich in organic carbon, available nutrients and physicochemical properties followed by pasture, horticulture, agriculture and wasteland, indicating the need for employing integrated soil fertility management in sustainable manner to optimize and maintain the soil health and favourable fertility status.

Impact

Soil testing and mapping provides information regarding physical health and nutrient availability in soil which forms the basic for the fertilizer recommendations and for thorough monitoring of the soil health for present and future agriculture.

26. Effect of Integrated Use of Biochar and Organic Manures on Soil Properties and Growth of Knol-Khol (*Brassica oleracea* var. *gongyloides*)

***Contact Scientist:* Dr. Piyush Sharma**

Objectives

- To study the effect of biochar and organic manures on soil properties.
- To examine the effect of biochar and organic manures on crop performance

Introduction

In India, about 93 million tons of crop residues are burned in each year. Residue burning traditionally provides a fast way to clear the agricultural field of residual biomass, facilitating further land preparation and planting. However, in addition to loss of valuable biomass and nutrients, biomass burning leads to release of toxic gases including GHGs. Studies sponsored by the Ministry of New and Renewable Energy (MNRE), Govt. of India have estimated surplus biomass availability at about 120–150 million tons/annum. Addition of organic matter to soil (*e.g.* manures, composts, sewage sludge) is common practice in farming systems to overcome carbon loss due to agricultural soils. Due to their low stability in the environment, these organic matter amendments require continued repeated application. The Biochar developed from these wastes can be added to soil to increase carbon sequestration and to improve crop yields and soil health. The knolkhol (Brassica oleracea var. gongylodes) cultivation in India is done in almost all parts of the country in small pockets having large area having production about 70,000 Mt.

Methodology

A field experiment was carried out at Chatha farm, SKUAST-Jammu. Treatments consists of T1. Control, T2. 100 per cent N through FYM.T3. 100 per cent N through vermicompost, T4. 50 per cent N through FYM+50 per cent N through vermicompost, T5. 2t/ha biochar, T6. 2t/ha biochar + T2, T7. 2t/ha biochar + T3, T8. 2t/ha biochar + T4, replicated three times in RBD.

Findings

Results revealed that addition of biochar with organic manures increased soil organic carbon, infiltration rate (IR) and decreased bulk density (BD) (Table 1). Highest soil infiltration rate (1.5 times) was recorded with the use of 2t/ha biochar and 50 per cent N through FYM+50 per cent N through vermicompost (T8), when compared to control. The higher value soil water retention and decreased BD may be attributed to porous structure and high surface area of biochar. The highest grain yield of knol khol (229 Q/ha) was recorded with the application of 50 per cent N through FYM +50 per cent N through vermicompost + 2t/ha biochar (T8) and was about 1.3 times higher over the control.

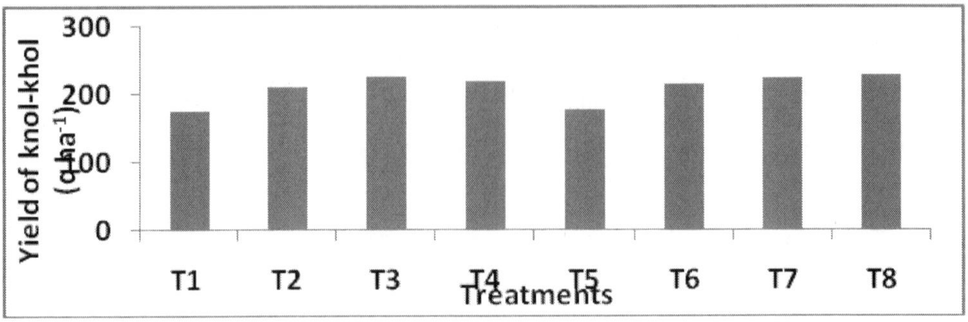

Table 1: Soil Properties as Influenced by Organic Manures and Biochar at Harvesting of Knol-khol

Treatments	*Bulk Density (Mg/m^3)*	*Infiltration Rate (mm/day)*
T1	1.39	204
T2	1.34	219
T3	1.33	220
T4	1.32	224
T5	1.33	215
T6	1.32	229
T7	1.32	300
T8	1.3	302
CD (0.05)	0.01	4.5

T1. Control, T2. 100 per cent N through FYM, T3. 100 per cent N through vermicompost, T4. 50 per cent N through FYM+50 per cent N through vermicompost, T5. 2t/ha biochar, T6. 2t/ha biochar +T2, T7. 2t/ha biochar + T3, T8. 2t/ha biochar +T4

Recommendation

Integration of 50 per cent N through FYM +50 per cent N through vermicompost + 2t/ha biochar under irrigated condition not only increased the yield of knol khol (1.3 times) over the control besides improving soil physical properties (soil moisture, infiltration rate, bulk density).

Impact

Application of biochar with different organic amendments has increased soil fertility, enhanced soil physical health and reduced the amount of carbon produced due to biomass burning.

27. Impact of Biochar Application on Soil Physical Properties, Soil Organic Carbon Dynamics and Crop Performance in Rice-Wheat System

***Contact Scientist:* Dr. Piyush Sharma**

Objective

To study the effect of organic manure management on soil properties and yield of rice

Introduction

Food demand and soil sustainability have become urgent concerns because of the impacts of global climate change. How to maintain the soil organic matter (SOM) levels in soils is a key consideration in agricultural productivity and carbon sequestration, particularly in agricultural lands in subtropical and tropical regions. Organic matter affects crops growth and yield either directly by supplying nutrients or indirectly by modifying soil physical properties such as stability of aggregates, porosity and available water capacity that can improve the root environment and stimulate plant growth. The added organic matters were usually mineralized to CO_2 rapidly leading to large-scale leakage in subtropical/tropical regions. Therefore, developing strategies for reducing the mineralization of added OM and increasing carbon sequestration in subtropical/tropical rural soils is necessary to facilitate land sustainability. Recently, co-application of biochar and other organic amendments has been determined to be an effective management practice for compensating for the limitations of applying biochar or an organic amendment alone. Therefore, the aim of this study was to find optimum biochar and compost amounts and mixture ratios with respect to plant response and soil properties.

Methodology

A field experiment was carried out at Chatha farm, SKUAST-Jammu. Treatments consists of T1: Control, T2: 100 per cent N through FYM, T3: 100 per cent N through vermicompost, T4: 50 per cent N through FYM +50 per cent N through vermicompost, T5: 1t/ha biochar (rice husk) +T2, T6: 1t/ha biochar (rice husk) + T3, T7: 1t/ha biochar (rice husk) + T4, T8: 2t/ha biochar (rice husk) + T2, T9: 2t/ha biochar (rice husk) + T3, T10: 2t/ha biochar (rice husk) + T4;replicated three times in RBD.

Findings

- Integrated nutrient management (50 per cent N through FYM +50 per cent N through vermicompost + 2t/ha biochar) recorded highest grain yield of basmati-370 and was about 83 per cent higher over the control under irrigated condition. Application of 50 per cent N through FYM +50 per cent N through vermicompost + 2t/ha biochar gave higher value of soil moisture (1.5 times), decreased soil compaction (1.1 times) and crack volume (1.8 times) in rice crop. As such use of 50 per cent N through FYM +50 per cent N through vermicompost + 2t/ha biochar is recommended in rice under irrigated condition.

- Less brown spot disease incidence was found in basmati-370 with the application of biochar in combination with FYM and Vermicompost as compare to control.

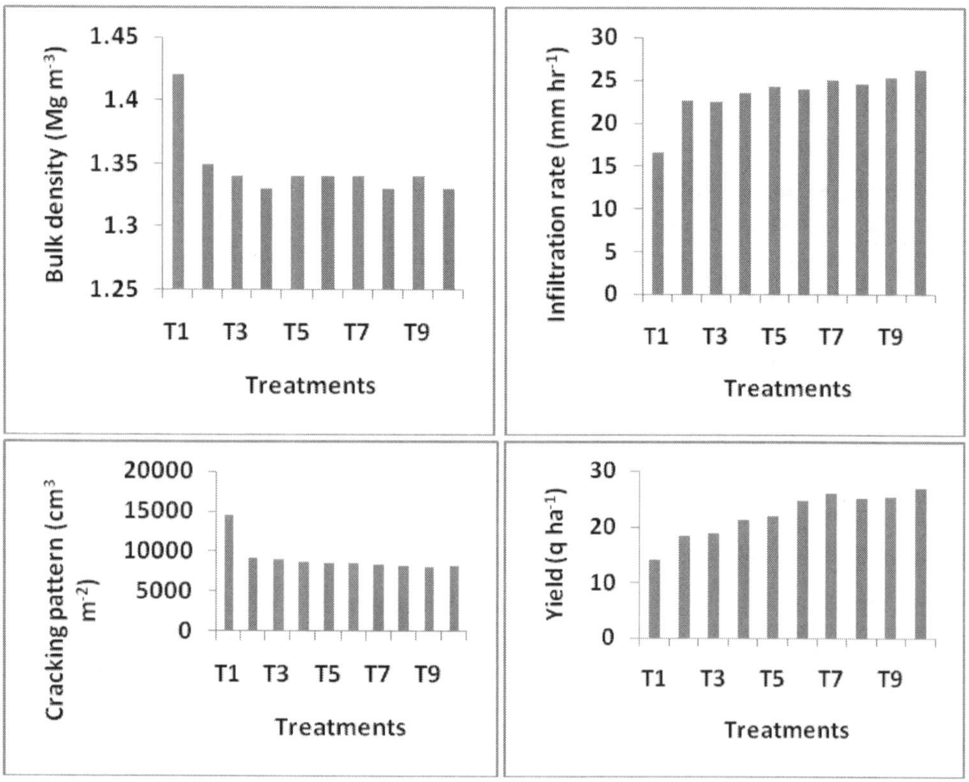

Rice Yield and Soil Properties as Affected by Organic Manures and Biochar.

Recommendation

Under irrigated condition, application of 50 per cent N through FYM +50 per cent N through vermicompost + 2t/ha biochar can help in reducing compaction, conserving soil moisture in puddled rice crop.

Impact

Application of biochar with different organic amendments has increased soil fertility, enhanced soil physical health and reduced the amount of carbon produced due to biomass burning.

28. Effects of Phosphorus and Sulphur Interaction on Nutrient Uptake and Yield of Black Gram (*Vigna mungo* L. Hepper)

Contact Scientist: Dr. A.P. Rai

The interaction of these nutrient elements may affect the critical levels of available P and S below which response to their application could be observed. Information on effect of combined application of P and S on yield, quality and content of each nutrient in black gram is rather limited in the sub-tropical zone of Jammu division of J&K. It is very much essential to develop a strong workable and compatible package of phosphorus and sulphur management for black gram based on scientific facts and local conditions.

Objective

Interaction effect of nutrient P and S on the growth and Yield of black gram

Methodology

The experiment was conducted to find the response of P and S in a factorial randomized block design with 3 replications comprised of 12 treatments. The treatments involved combination of P @ 0, 15, 30 and 45 kg ha^{-1} and S @ 0, 15 and 30 kg ha^{-1}. Doses of S were 0, 20, 40, kg ha^{-1} and three levels of Mo *viz.*, 0, 0.5, 1.0kg ha^{-1} were applied through soil application. The initial Surface (0-15cm) soil samples and after harvesting soil sample were collected and analyzed for following the method of Jackson (1973).

Results

Table 1: Effect of different Levels of Phosphorus and Sulphur Application on Seed Yield (kg ha^{-1}) of Black Gram (cv. Uttara)

Phosphorus Levels (kg ha^{-1})	*Sulphur Levels (kg ha^{-1})*			*Mean*
	0	*15*	*30*	
0	466.56	501.20	530.50	**499.42**
15	550.50	645.02	715.55	**637.02**
30	620.00	740.25	859.53	**739.93**
45	783.50	880.50	955.50	**873.17**
Mean	**605.14**	**691.74**	**765.27**	
CD (P = 0.05) P levels: 30.75, S levels: 25.15, P x S: 25.25				

Recommendation

Application of 45 kg P ha^{-1} and 30 kg S ha^{-1} at the sowing time increased grain yield of black gram 61 and 26 percent over control respectively. Application of P and S also increased available P in soil as compared to control.

Impact

Application of P and S also increased available P in soil as compared to control. The technology will enhance crop production thereby resulting in increased crop productivity.

29. Response of Zinc and Iron Application in Rice (Cv. Basmati-370)

Contact Scientist: **Dr. A.P. Rai**

Objective

Effect of zinc and iron application to Basmati-370

Methodology

The experiment was conducted to find response of Zn fertilization to Basmati rice '370'in a factorial randomized block design with 3 replications comprised of 12 treatments [Soil application of $ZnSO_4.7H_2O$ @ 0, 5, 10 and 20 kg ha^{-1} and $FeSO_4.7H_2O$ @ 0, 2.5 and 5.0 kg $ha^{-1}$and foliar application of 0.2 per cent $ZnSO_4.7H_2O$] at tillering and flowering stage during *Kharif* seasons of 2017.

Findings

See Table 1.

Recommendation

Application of Zn @ 20 kg ha^{-1} and Fe @ 5.0 kg ha^{-1} along with foliar application of Zn and Fe content significant increased Zn (17.80 per cent) and Fe (18.96 per cent) in grain, and Zn (16.44 per cent) and Fe (20.47 per cent) straw.

Impact

Application of Zn and Fe enhance the soil fertility as well as quality of rice (enrich with Zn and Fe).

Table 1: Effect of Zn and Fe Levels on the Zn and Fe Concentration in Grain and Straw of Basmati Rice

Levels	*Zn Concentration (mg kg^{-1}) in Basmati Rice*									
	Grain					*Straw*				
	Zn (kg ha^{-1})					*Zn (kg ha^{-1})*				
Fe (kg ha^{-1})	*0*	*5*	*10*	*20*	*Mean*	*0*	*5*	*10*	*20*	*Mean*
0	10.9	11.32	11.52	12.61	11.59	12.34	13.32	14.35	13.82	13.46
2.5	12.64	13.21	14.21	14.36	13.61	13.35	14.54	15.61	15.82	14.83
5.0	12.34	13.64	14.89	15.31	14.05	14.64	15.61	16.61	17.32	16.05
Mean	11.96	12.72	13.54	14.09		13.44	14.49	15.52	15.65	
CD (P = 0.05)	Zn= 0.81	Fe=0.71	ZnxFe=NS			Zn= 1.4	Fe=1.2	ZnxFe=NS		

Levels	*Fe Concentration (mg kg^{-1}) in Basmati Rice*									
	Grain					*Straw*				
	Zn (kg ha^{-1})					*Zn (kg ha^{-1})*				
Fe (kg ha^{-1})	*0*	*5*	*10*	*20*	*Mean*	*0*	*5*	*10*	*20*	*Mean*
0	9.64	10.54	10.64	9.64	10.12	10.64	11.54	12.51	12.34	12.13
2.5	11.32	13.51	14.61	15.34	13.70	14.35	15.64	17.64	18.61	17.30
5.0	16.54	17.61	17.68	19.62	17.86	19.64	20.13	21.35	22.34	21.27
Mean	12.50	13.89	14.31	14.87		17.00	17.89	19.50	20.48	
CD (P = 0.05)	Zn= 0.41	Fe=0.91	ZnxFe=NS			Zn= 1.6	Fe=1.4	ZnxFe=NS		

30. Management of Mustard Crop through Sulphur and Zinc Application

Contact Scientist: Dr. A.P. Rai

Sulphur has also important role on reducing male sterility. Mustard in general is very sensitive to Sulphur deficiency. Therefore, site specific Sulphur and Znic management through soil test recommendation based should be adopted to improve upon the yield and yield attributes in mustard crop under Jammu, where deficiency of Sulphur and Zinc has already been reported.

Objective

Interaction effect of nutrient S, and Mo on the growth, Yield and quality

Methodology

The treatments consisted of Four levels of sulphur (0, 10, 20 and 30 kg S ha^{-1}) and three levels of Zinc (0, 5 and 10 kg Zn ha^{-1}) which were replicated three times in a factorial RBD design. Full doses of Sulphur and Zinc were applied as per treatment at the time of sowing in the form of gypsum and $ZnSO_4.7H_2O$, respectively. Sulphur was analyzed by turbidimetric method (Chesnin and Yien 1951). Zinc was determined by 0.005M DTPA, 0.01M $CaCl_2$ and 0.1 M Triethanolamine (pH 7.3) as per the procedure described by Lindsay and Norvell (1978).

Table 1: Effect of different Levels of Sulphur and Zinc Application on Seed Yield and Oil Content of Mustard (Cv. RSPR 01)

Seed Yield				
Sulphur Levels (kg ha^{-1})	*Zinc Levels (kg ha^{-1})*			*Mean*
	0	*5*	*10*	
0	830	920	980	977
10	970	992	1013	1066
20	1062	1110	1180	1116
30	1124	1154	1162	1109
Mean	996.50	1044.0	1083.75	
CD (P = 0.05) P levels: 41, S levels: 34, P x S: 53				

Oil Content (per cent)				
Sulphur Levels (kg ha^{-1})	*Zinc Levels (kg ha^{-1})*			*Mean*
	0	*5*	*10*	
10	33.42	35.13	35.26	**34.60**
20	34.82	36.37	37.14	**36.11**
30	35.92	36.54	36.92	**36.46**
Mean	**34.31**	**35.54**	**35.93**	
CD (P = 0.05) P levels: 34, S levels: 0.58, P x S: 1.13				

Recommendation

Application of 20kg S and 10 Kg Zn ha^{-1} gives the highest values of seed yield 1180 kg/ha and 11.54 per cent increase oil content in the mustard over the control.

Impact

This technology provide for application of S and Zn in better mustard crop production.

31. Training and Demonstration of LCC for Real Time Nitrogen Management under Climate Smart Agriculture Practices (CSAP) for Maintaining Ecological Integrity through KVK'S

***Contact Scientist:* Dr. Vivak M. Arya**

Objective

1. Training and resource development
2. Demonstration at farmers field
3. Popularization of LCC under CSAP in the nearby villages through awareness programmes

Introduction

The leaf color chart (LCC) is an easy-to use and inexpensive diagnostic tool for monitoring the relative greenness of a rice leaf as an indicator of the plant N status. The LCC can be used to guide the application of fertilizer N to maintain optimal leaf N content for achieving high rice yield with effective N management.

Research Methodology

Use LCC for need based nitrogen application to the paddy. Use of LCC can save excessive use of nitrogen which is now a day's polluting the soil as well as water. Apply 60kg/ha of Urea as a basal dose after the completion of puddling operation.

Match the colour of the youngest fully expanded leaf without any disease symptoms (2nd from top) of at least 10randomly selected free rice plants with the colour strip of the leaf colour chart every 7-10 days starting from 2 weeks after transplanting of paddy til the flowering stage. Every time while matching the leaves with the strips of LCC if the greenness stage. Every time while matching the leaves is less than the shades '4' on the leaf colour strip, top dress 75kg Urea/ha. If the colourof 5 or more out of 10 leaves is greener than the shade '4' on the LCC strip, do not apply any urea to paddy grown in all types of soil. Even if FYM is used use of LCC gives good result in saving the excessive use of nitrogen.

Methodology Adopted for Transferring the Technology to Farmers

1. Selection of Districts (Jammu,Samba, Kathua and Udhampur)
2. From each district three villages are selected (4 X 2= 8)
3. Awareness Camps/trainings for the farmers (10)
4. Training/Interaction in between the crop season
5. Data collection from the adopted fields
6. Generation of base line data for comparison between adopted and non-adopted farmers
7. Distribution of LCC on large scale to the farmers

Results

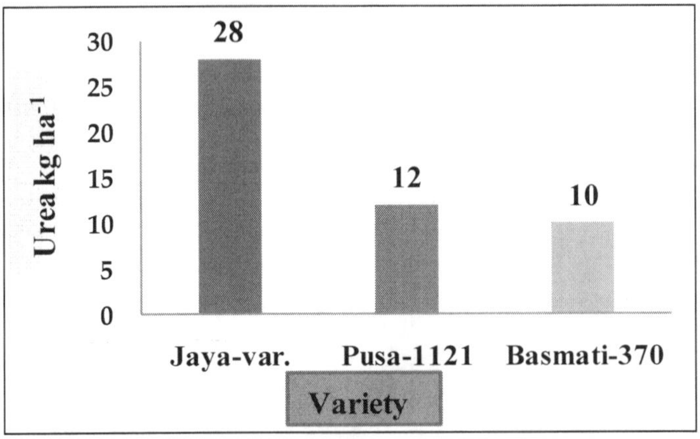

Usage of LCC (Leaf Colour chart) Saves Urea (kgha^{-1}) in Jaya-var., Pusa-1121 and Basmati-3 of Various Paddy Varieties under LCC and Farmer Practices.

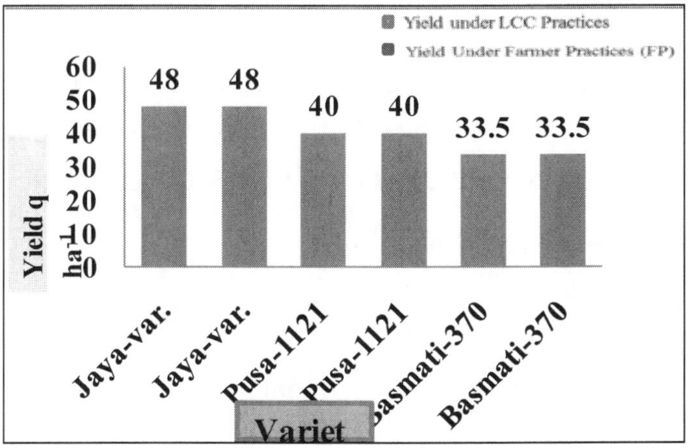

Yield (qha^{-1})

Recommendations

- ❒ Use of LCC can save 26-28 kg of urea/ha in coarse variety and 10-12 kg of urea/ha in Basmati

32. Assessment of Fertilizer Recommendations in Maize-Wheat, based on Target Yield Concept (TYC) under Subtropical Plains of Jammu

***Contact Scientist:* Dr. Vivak M. Arya**

Objectives

1. Validation of fertilizer adjustment equations for Maize-wheat in subtropical plains of Jammu
2. Formulation of fertilizer recommendations based on TYC.
3. To enhance soil productivity and nutrient use efficiency through targeted yield approach

Introduction

A field experiment was conducted at KVK- Kathua research farm of Sher-e-Kashmir University of Agricultural Sciences and Technology of Jammu to study the fertilizer recommendations in Maize-wheat based on target yield concept during *Kharif* and *Rabi* 2015. The experiment was laid out in RBD with maize variety Double Dekalb and wheat variety HD 2967. Organic source of fertilizers was supplied through FYM and inorganic dose was given by urea, DAP and MOP. The soil of experimental field was sandy clay loam in texture, slightly alkaline in reaction (pH 7.03), low in organic carbon (4.8 g/kg) and available nitrogen (185 kg/ha) and medium in available phosphorus (18 kg/ha) and low in available K (95 kg/ha).

Treatments

1. T_1: Control
2. T_2: Recommended dose of Fertilizers (100 per cent inorganics)
3. T_3: RDF (75 per cent inorganics + 25 per cent organics)
4. T_4: Soil test based fertilizer dose (100 per cent inorganics)
5. T_5: Soil test based fertilizer dose (75 per cent inorganics + 25 per cent organics)
6. $T_{6:}$ Target Yield based fertilizer dose (100 per cent inorganics)
7. T_7: Target Yield based fertilizer dose (75 per cent inorganics + 25 per cent organics)

Recommendations

1. Fertilizer recommendation based on target yield (75 per cent

Inorganics+25 per cent organic) gave the highest yield of 55 q/ha of maize in irrigated conditions of Jammu plains. As such **use of integrated nutrient management practices is recommended in** maize in irrigated conditions of Jammu plains.

2. Fertilizer recommendation based on target yield (75 per cent Inorganics+25 per cent organics) gave the highest yield of 45 q/ha of wheat in irrigated conditions of Jammu plains. As such **use of integrated nutrient management practices is recommended in** wheat in irrigated conditions of Jammu plain.

Division of Agricultural Economics and Agribusiness Management

33. Economics of Niche Crops and Assessment of Farm Level Diversification in Jammu Region

Contact Scientist: **Dr. Sudhakar Dwivedi**

Objective(s)

- To work out the economics of niche crop *viz.* Basmati rice

Methodology

The present study was conducted in the Jammu division of the Jammu and Kashmir State. The district of Jammu region namely; Jammu was selected purposively for assessing the economics of Basmati rice. The primary data regarding capital investment, resource use and cost and return of basmati rice crop was collected using multi-stage random sampling method. In the second stage of sampling, two major sub-divisions of Jammu district in terms of production of basmati rice namely; the sub-division R.S. Pura and the sub-division Marh were selected. In the third stage, one major basmati producing block from each sub-division *i.e.* R.S. Pura block from R.S. Pura sub-division and Marh block from Marh sub-division were selected. In the fourth stage, five villages from each chosen block were selected randomly namely; Baga Jana, Jindlehar, Jinder-Melu, Chohalla and

Gagian from R.S. Pura block and Gajansoo, Kanachak, Kalyanpur, pounichak and Gho Manasa from Marh block. A list of Basmati growers in each of these ten villages was prepared and ten farmers from each village were again selected randomly without replacement. Thus in total, a sample size of hundred farmers was drawn for collection of data on basmati rice crop grown in Kharif season of year 2013.

Data/Table

Table 1: Cost-Benefit Analysis of Basmati Production (per acre)

Cost Structure		*Productivity and Returns*			
Particulars	*Rs.*	*Particulars*	*Quantity (quintals)*	*Price (Rs.)*	*Income (Rs.)*
Seed	496.00	**Main Product**	12.61	5794.88	60463.44
Irrigation	1442.76	**By-product**	20.75	424.76	8815.21
Manure and Fertilizers	827.43	**Gross Return**			69278.65
Plant Protection Chemicals	339.42	**Net Return**			42939.72
Machinery and Human Labour	10836.61	**C.B ratio**			1:2.63
Interest on working capital @ 14 per cent	1951.91				
Total variable cost	15894.13				
Land rent	8923.55				
Depreciation on Tractor	267.65				
Depreciation on Electric motor	73.02				
Depreciation on Intercultural implements	61.48				
Interest on fixed cost @ 12 per cent	1119.08				
Total Fixed Cost	10444.80				
Total cost	**26338.93**				

Recommendations

The price of basmati crop needs regulation to build confidence among the growers to produce high quality unique crops in Jammu region.

Impact

The results are helpful for policy makers, farming community as well as other researchers/academicians in this field.

34. An Economic Analysis of Production and Marketing of Rajmash Crop in Doda District

Contact Scientist: **Dr. S.P. Singh**

Objectives

- To analyze the costs and returns of Rajmash production

Methodology

Locale of study: Doda district of J&K state

Period of the study: 2015-16

Database: Both Primary and Secondary data used. Primary data was collected using personal interview method with the help of an especially prepared and pre-tested schedule/questionnaire.

Sampling Design: Multistage sampling technique was used.

Sampling Structure

1. **Selection of blocks**: Out of 17 blocks, 02 blocks *viz.*: Bhaderwah and Bhalla blocks were selected as they had highest no. of Rajmash cultivators.
2. **Selection of villages:** 05 villages from each block were selected randomly.
3. **Selection of farmers**: 10 Farming households were selected randomly from each village. Overall, a total sample size consisted of 100 farming households.

Categorization of farmers: According to their land holdings, farmers were categorized into Marginal (0-1 hectare), Small (> 1 to <2 hectare) and Medium (>2 hectares)

Cost and returns Analysis: For computation of costs and returns, the concepts framed by CACP were used like Cost A, B, C *etc.*

Table 1: Costs and Returns of Rajmash on Sampled Farms under Study (₹/ha)

Sl.No.	*Particulars*	*Marginal*	*Small*	*Medium*	*All farms*
1.	Cost				
	Total variable cost	20092.78	20636.05	21160.31	20481.70
	Total fixed cost	13623.32	15664.69	16770.66	14872.30
	Total cost	33716.10	36300.74	37930.97	35354.00

Sl.No.	Particulars	Marginal	Small	Medium	All farms
2.	Returns				
	Gross returns	71477.00	70597.97	70463.83	70987.93
	Net returns	37760.90	34297.23	32532.86	35633.93
3.	Cost-Benefit Ratio				
	C-B Ratio on Cost C_2	1 : 2.12	1 : 1.94	1 : 1.86	1 : 2.01

Recommendations

- Rajmash being a remunerative crop suiting to the topographical conditions of the hilly/mountainous regions should be encouraged as an inter-crop with maize and farmers needs to be supported with financial assistance, good quality seed and technology.
- There is a possibility of more income for medium and small farmers, if they learn from marginal farmers the art of making use of resources more efficiently.

Impact/Benefits

- Useful for Policy makers, Financial institutions, Researchers and progressive farmers in their respective fields (as no such ground level studies were available in the past

35. Economic Analysis of Production and Marketing of Marigold Flowers in Jammu Subtropics

***Contact Scientist:* Dr. Anil Bhat**

Objectives

1. To find out the marketing channels and price spread of marigold flower
2. To identify opportunities and constraints in the production and marketing of marigold flower

Methodologies

A multistage random sampling technique was followed for selection of marigold farmers. Two districts Jammu and Kathua were selected purposively from the Jammu subtropics as these districts had highest area under cultivation during 2014-15 and five villages were selected from each district on the basis of highest number of marigold farmers, so as to constitute a total of 10 villages. Ten farmers each were selected randomly from the selected villages to make a sample size of 100 farmers in all which were surveyed with the help of pre tested questionnaire. Along with the farmers, wholesalers and retailers from the study area were also surveyed for data regarding marketing.

Cost-Benefit Analysis of Production

For Estimating the Costs and Returns

The cost and returns analysis was worked out using CACP cost concepts like Cost A_1, cost A_2, Cost B_1, cost B_2, Cost C_1, Cost C_2 and Cost C_3

Analysis of Marketing

The data collected was tabulated and analyzed for examining the marketing cost, margins, price spread and the marketing efficiency using the formula given by Acharya and Agarwal (2001).

Data/Table

Table 1: Marketing Analysis of different Channels for Marigold Flowers in Jammu District

Particulars	*Farmer-Retailer-Consumer*	*Farmer-Wholesaler-Retailer-Consumer*	*Farmer-Retailer*-Consumer*
Net price received by the farmer (₹/kg)	22.70	20.30	24.10
Marketing margin (₹/kg)	19.65	18.68	17.49

Particulars	Farmer-Retailer-Consumer	Farmer-Wholesaler-Retailer-Consumer	Farmer-Retailer*-Consumer
Marketing cost (₹/kg)	9.00	10.50	8.50
Marketing loss (₹/kg)	8.65	9.52	6.91
Marketing efficiency	0.61	0.52	0.73

Table 2: Marketing Analysis of different Channels for Marigold Flowers in Kathua District

Particulars	Farmer-Retailer-Consumer	Farmer-Wholesaler-Retailer-Consumer	Farmer-Retailer*-Consumer	Farmer-Retailer**-Consumer	Farmer-Consumer
Net price received by the farmer (₹/kg)	32.70	27.30	36.80	37.80	52.60
Marketing margin (₹/kg)	20.82	18.68	19.30	19.50	0.00
Marketing cost (₹/kg)	9.00	14.50	8.00	8.00	12.00
Marketing loss (₹/kg)	7.48	9.52	5.90	4.70	5.40
Marketing efficiency	0.87	0.64	1.11	1.17	3.02

Table 3: Comparison of Employment Generation and Benefits between Marigold and Cereal Crops

Sl. No.	Operations	Marigold		Rice		Wheat	
		Family Labour (nos)	Hired Labour (nos)	Family Labour (nos)	Hired Labour (nos)	Family Labour (nos)	Hired Labour (nos)
1	Land preparation	6.00	4.00	10.00	5.00	4.00	4.00
2	Nursery raising/ Sowing	3.06	0.00	1.51	0.00	1.00	0.00
3	Transplanting	9.66	6.67	6.38	12.73	0.00	0.00
4	Fertilizer application	2.00	0.00	1.21	0.00	3.00	0.00
5	Irrigation	27.54	0.00	20.25	0.00	10.29	0.00

Sl. No.	*Operations*	*Marigold*		*Rice*		*Wheat*	
		Family Labour (nos)	*Hired Labour (nos)*	*Family Labour (nos)*	*Hired Labour (nos)*	*Family Labour (nos)*	*Hired Labour (nos)*
6	Intercultural operations	17.30	12.36	0.00	0.00	0.00	0.00
7	Pesticide application	6.25	0.00	3.29	0.00	3.29	0.00
8	Harvesting and threshing	30.00	0.00	10.00	15.00	10.00	14.00
9	Total	101.81	23.03	52.64	32.73	31.58	18.00
10	Production (main) (q) (byproduct) (q)	168.09 0.00		23.94 36.86		19.78 12.84	
11	Gross income (main) (₹) (byproduct) (₹)	592440.30 0.00		102216.394054.97		31642.40 10273.52	
12	Gross returns (₹)	592440.30		106271.36		41915.92	
13	Cost of cultivation (₹)	117697.70		40359.01		22730.69	
14	Net returns (₹)	474742.60		65912.35		19195.23	
15	BC ratio	1:5.03		1:2.63		1:1.84	

Recommendations

- The farmers of study area are recommended for shifting the part of land from traditional cereal crop to marigold cultivation as it will provide them regular income but before shifting, they should plan the marketing aspect well in advance.
- The farmers are recommended that they should focus on value addition rather than production aspect as by making garlands, their income will be increased by more than 40 percent.
- The farmers are suggested to focus on marigold crop cultivation because it will increase their employment opportunities by increasing their labour requirement by more than 30 percent.

Benefits

- Cost and return analysis of marigold cultivation showed that marigold crop has the potential of generating gross returns of ₹5 per rupee invested.
- A comparison of employment opportunities revealed that marigold crop is a labour intensive crop and labour requirement of the farmer increases by

30 percent on cultivating marigold on same piece of land when compared to cereals.

- Marigold crop has the potential of generating ₹5 per rupee invested while cereals crops namely rice and wheat generated gross returns of ₹2.63 and ₹1.84, respectively.

Division of Food Technology

36. Development of Fiber Rich Oat-Wheat Flour Blended Biscuits using Mango Peel

***Contact Scientist:* Dr. Jagmohan Singh**

Objectives

1. To analyze the chemical characteristics of raw material to be used.
2. To determine the effect of blending of oats on physico-chemical, microbial and sensory evaluation of developed product.
3. To estimate the shelf life and economics of the developed product.

Introduction

Mango (*Mangifera indica* L.) is one of the most important fruits worldwide and is cultivated in more than 100 countries at both tropical and subtropical latitude, especially in Asia. The mango mainly constitutes pulp 33-70 per cent followed by kernel 7-24 per cent and peel 15-20 per cent of the fruit weight. Its peel is a promising source of phytonutrient such as polyphenols, carotenoids and Vitamin E and C. Higher polyphenol contents are present in mango peel than that of pulp. Intake of dietary fiber and phytochemical such as polyphenols, carotenoids, tocopherols and ascorbic acid have been related to the maintenance of health and protection from diseases such as cancer, cardiovascular diseases and many other degenerative diseases.

Oats (*Avena Sativa* L.) ranks sixth around the world cereals production statistics following wheat, maize, rice, barley and sorghum. The amount of oats used for human consumption has increased progressively, the fact health effects of oats benefits mainly on the dietary fibre and β-glucan content. Wheat (*Triticum aestivium* L.) is the major contributor of protein content of daily diet. In the recent years, an interest in the bakery products, with increased nutritional value, such as fibre enriched products has increased Biscuits are the most popularly consumed bakery items in the world. Keeping in view, present study was undertaken for the development of fiber rich oat-wheat flour blended biscuits using mango peel powders of two cultivars *viz.* Chausa and Langra at different levels to improve the nutraceutical properties of the developed product.

Methodology

Fibre rich oat-wheat flour blended biscuits using mango peel were prepared by using Wheat flour(WF), Oat flour(OF) and mango peel powder(MPP) of two cultivar *viz.* Chausa (CMPP) and Langra (LMPP) at different replacing levels in the ratios of T_1(100:0:0::WF:OF:MPP); T_2(0:100:0:: WF:OF:MPP); T_3(50:42:8::WF:OF:CMPP); T_4(50:42:8::WF:OF:LMPP); T_5(50:40:10::WF:OF: CMPP) ; T_6(50:40:10::WF:OF:LMPP); T_7(50:38:12::WF:OF:CMPP); T_8(50:38:12::WF:OF: LMPP); T_9(50:36:14::WF:OF:CMPP); T_{10}(50:36:14::WF:OF:LMPP); T_{11}(50:34:16::WF:OF: CMPP) and T_{12}(50:34:16::WF:OF:LMPP).

Flow Chart for the Preparation of Fibre Rich Biscuits

Mixing of Wheat flour + Oat flour + Mango Peel Powder

↓

Creaming of sugar and fat

↓

Mixing of all the ingredients

↓

Kneading into Dough

↓

Rolling into sheets

↓

Cutting with a circular mould

↓

Baking at 205°C for 10 min

↓

Cooling at room temperature

↓

Packaging in pouches

↓

Storage at room temperature

The processed product was stored at ambient conditions and subjected to chemical, microbial and sensory evaluations at an interval of one month for a period of three months.

Table 1: Effect of Treatments and Storage Period on Overall Acceptability Score of Fibre Rich Biscuits

Treatments	*Storage Period (Days)*				*Mean*
	0	*30*	*60*	*90*	
T1 (100:0:0::WF:OF:MPP)	8.08	8.02	7.83	7.55	7.87
T2 (0:100:0::WF:OF:MPP)	7.59	7.48	7.32	7.12	7.38
T3 (50:42:8::WF:OF:CMPP)	7.73	7.65	7.51	7.32	7.55
T4 (50:42:8::WF:OF:LMPP)	7.68	7.55	7.43	7.24	7.48
T5 (50:40:10::WF:OF:CMPP)	7.78	7.71	7.53	7.29	7.58
T6 (50:40:10::WF:OF:LMPP)	7.73	7.62	7.42	7.25	7.50
T7 (50:38:12::WF:OF:CMPP)	7.51	7.43	7.23	7.04	7.30
T8 (50:38:12::WF:OF:LMPP)	7.43	7.33	7.19	6.99	7.23
T9 (50:36:14::WF:OF:CMPP)	7.23	7.15	6.97	6.76	7.03
T10 (50:36:14::WF:OF:LMPP)	7.19	7.07	6.92	6.66	6.96
T11 (50:34:16::WF:OF:CMPP)	6.97	6.85	6.73	6.50	6.76
T12 (50:34:16::WF:OF:LMPP)	6.93	6.79	6.64	6.42	6.70
	7.49	7.39	7.23	7.01	

Factors C.D. (p=0.05) Treatment 0.01

Storage 0.01 Treatment x Storage 0.02

WF: Wheat Flour **OF**: Oat flour **MPP:** Mango Peel Powder **CMPP:** Chausa Mango Peel Powder; **LMPP:** Langra Mango Peel Powder

Recommendation

Fibre rich biscuits developed by blending 50 per cent wheat flour, 40 per cent oat flour and 10 per cent mango peel powder had highest level of acceptance. It had crude fibre of 4.55 per cent, total phenolic content of 4.16 mg GAE/g, scavenging activity of 29.84 per cent and microbial count of 3.43×10^{2} c.f.u/g during three months of storage. Cost of biscuits was found to be Rs 14.97/100g which is cheaper than commercially available biscuit *i.e.* Rs 20/100g.

Preparation of Biscuits

Fibre Rich Biscuits

Benefits

With the growing interest of consumers in health–related foods, the addition of oat and mango peel powder to wheat flour affected the nutrition composition and sensory characteristics in various ways. Oat and mango peel powder can be used as a suitable source of dietary fiber with associated bioactive compounds and can be incorporated as ingredient in large variety of food products.

37. Effect of Pre-treatments and Dehydration on the Quality of Oyster Mushroom

Contact Scientist: **Dr. Anju Bhat**

Objectives

1. To study the effect of pre treatments and drying methods on the quality of dehydrated oyster mushrooms.
2. To study the effect of packaging and storage on the shelf life of dehydrated oyster.

Introduction

Mushrooms assume considerable importance in the human diet as they are rich in non-starchy carbohydrates, dietary fibre, minerals, vitamin-B and low in fat content. In India, there are mainly three species of mushrooms, namely, white button mushroom (*Agaricus bisporus*), oyster mushrooms (*Pleurotus sajor caju*) and paddy straw mushrooms (*Volvariella valvacea*), that are grown commercially. Among these, oyster mushrooms (*Pleurotus sajor caju*) possess unique nutritional and medicinal values, characteristic aroma and taste Oyster mushrooms are suitable for cultivation in plains and even in hilly areas. They contain 89.8 per cent moisture, 2.9 per cent protein, 0.36 per cent fat, 5.3 per cent carbohydrates, 1 per cent fiber, 1 per cent ash and 33.9 Kilocalories of energy. Mushrooms have become attractive as a functional food and as a source for the development of drugs and nutraceuticals (Ferreira *et al.*, , 2009) due to their antioxidants antitumor and antimicrobial properties. Oyster mushrooms are perishable, because they have a high respiration and transpiration rates leading to rapid post-harvest deterioration. Browning reactions and dehydration affect the shelf-life of fresh oyster mushrooms within a few days, even if refrigerated.

Drying is a method of preservation in which the water activity of the food is reduced. Conventional hot-air drying is one of the most frequently used methods for mushroom dehydration, which involves thermal or chemical pretreatment and drying at temperatures between 50 and 80 °C.

Methodology

1. Standardization of Drying Method and Temperature

Mushrooms were blanched and then subjected to drying by the following methods

Drying Method	*Temperature (°C)*
T_1 (Cabinet drying)	40
T_2 (Cabinet drying)	45
T_3 (Cabinet drying)	50
T_4 (Cabinet drying)	55
T_5 (Solar drying)	-
Microwave Drying	*Power Levels (W)*
T_6	300
T_7	450
T_8	600
T_9	Dehydration (Cabinet drying at 50°C) + microwave drying
T_{10}	Solar drying + microwave drying

The drying method standardized in this experiment was used to dry mushrooms in next experiment.

2. Effect of Pre-treatments on Drying of Oyster Mushrooms

Mushrooms were subjected to following pre-treatments for 15 minutes and then dried in cabinet dryer at temperature 50°C.

Pre-treatments	
T_1	Control
T_2	0.1 per cent KMS
T_3	0.2 per cent KMS
T_4	1 per cent CaCl2
T_5	2 per cent CaCl2
T_6	0.5 per cent Glycerol
T7	1.0 per cent Glycerol

The best pre-treatment was selected from above and mushrooms were packaged in LDPE,

PET jars, aluminium laminated pouches and stored at room temperature to assess storage life.

Table 1: Effect of Drying Method and Temperature on Moisture Content (per cent), Shrinkage Ratio, Dehydration Ratio, Rehydration Ratio, Drying Time (Minutes) and Water Activity (aw) of Oyster Mushrooms

Drying Method and Temperature	*Moisture Content (per cent)*	*Shrinkage Ratio*	*De-hydration Ratio*	*Re-hydration Ratio*	*Drying Time (minutes)*	*Water Activity*
T1 Cabinet drying at 40 °C	8.75	0.90	9.58	3.15	420	0.56
T2 Cabinet drying at 45 °C	8.72	0.92	9.55	3.12	390	0.55
T3 Cabinet drying at 50 °C	8.67	0.90	9.60	3.18	330	0.52
T4 Cabinet drying at 55 °C	8.70	0.94	9.48	3.07	350	0.54
T5 Solar drying	8.73	0.88	9.52	2.78	600	0.50
T6 Microwave drying at 300 W	8.60	0.84	9.40	2.74	120	0.46
T7 Microwave drying at 450 W	8.57	0.82	9.42	2.82	90	0.44
T8 Microwave drying at 600 W	8.54	0.80	9.46	2.90	70	0.43
T9 Cabinet drying (50°C) + microwave drying	8.62	0.85	9.50	2.69	240	0.47
T10 Solar drying + microwave drying	8.65	0.86	9.48	2.62	320	0.48
Mean	8.65	0.87	9.50	2.91	293	0.49
CD(0.05)	N.S.	0.04	0.07	0.12	27.7	N.S.
±S.E.(m)	0.02	0.01	0.02	0.04	9.33	0.03

Table 2: Effect of Drying Method and Temperature on Browning Index of Oyster Mushrooms

Drying Method and Temperature	*Browning Index*
T_1 Cabinet drying at 40 °C	0.10
T_2 Cabinet drying at 45 °C	0.11
T_3 Cabinet drying at 50 °C	0.09
T_4 Cabinet drying at 55 °C	0.13
T_5 Solar drying	0.20

Drying Method and Temperature	*Browning Index*
T_6 Microwave drying at 300 W	0.10
T_7 Microwave drying at 450 W	0.11
T_8 Microwave drying at 600 W	0.12
T_9 Cabinet drying (50 °C) +microwave drying	0.13
T_{10} Solar drying + microwavedrying	0.18
Mean	0.13
CD(0.05)	0.09
± S.E.(m)	**0.03**

Table 3: Effect of Pretreatments on Appearance, Flavour, Texture and Taste of Dried Oyster Mushrooms

Pre-treatments	*Appearance*	*Flavour*	*Texture*	*Taste*
T1 (Control)	7.08	6.84	6.96	6.81
T2 (0.1 per cent KMS)	8.43	8.05	8.15	8.03
T3 (0.2 per cent KMS)	8.47	8.10	8.18	8.10
T4 (1 per cent CaCl2)	8.40	7.90	8.22	7.91
T5 (2 per cent CaCl2)	8.44	7.97	8.28	7.96
T6 (0.5 per cent Glycerol)	8.56	8.19	8.35	8.21
T7 (1 per cent Glycerol)	8.62	8.25	8.44	8.27
Mean	8.28	7.90	8.08	7.89
CD (0.05)	0.07	0.07	0.08	0.07
±S.E.(m)	0.02	0.02	0.03	0.02

Table 4: Effect of Packaging and Storage on the Overall Acceptability of Pretreated (1 per cent glycerol) and Cabinet Dried Oyster Mushrooms

Packaging Material	*Overall Acceptability*				
	Storage Period (Days)				
	0	*30*	*60*	*90*	*Mean*
LDPE	8.33	8.05	7.90	7.70	7.99
PET Jars	8.33	8.15	8.00	7.80	8.07
Aluminium laminated pouches	8.33	8.20	8.10	8.00	8.16
Mean	8.33	8.13	8.00	7.83	
		CD(0.05)		**±S.E.(m)**	
Packaging material		0.05		0.02	
Storage		0.06		0.02	
Packaging material × Storage	0.10			0.03	

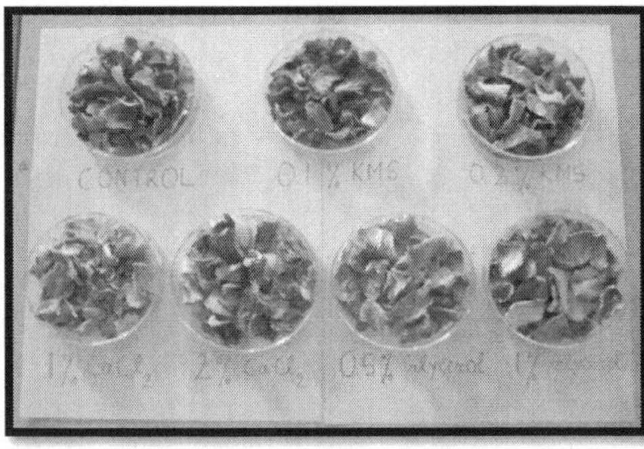

Dried Oyster Mushrooms **Packed Oyster Mushrooms**

Recommendation

Oyster mushrooms pre treated with 1 per cent glycerol and dried in cabinet tray drier at 50°C was found to be the best among other pre treatments due to low moisture content (8.30 per cent), water activity (0.44), microbial count (1.32 c.f.u/g), highest dehydration ratio (9.84), crude fat (1.88 per cent) and sensory attributes. Mushroom packaged in aluminium laminated pouches, enhanced shelf life upto 90 days of storage.

Benefits

Pre-treatment is necessary to check discolouration in mushroom and also helps in checking enzymatic browning, stabilizing colour, enhancing flavour retention and maintaining textural properties. Packing of dried mushroom in aluminium laminated pouches maintains quality during storage.

38. Development and Quality Evaluation of Brown Rice Based Weaning Food

***Contact Scientist:* Dr. Monica Sood**

Objectives

- To standardize weaning food from multipurpose flour (brown rice and mung bean), apple pulp powder and walnut.
- To ascertain the changes in the physico-chemical and sensory properties of prepared weaning food.

Introduction

The word "wean" is derived from the Anglo-Saxon "*wenian*" meaning "to accustom" (as a child) to take food otherwise than by nursing. The main concern is making sure that there is no gap between nutrient requirements and what a child is able to consume, absorb and utilize. Traditionally weaning foods are liquids and semi-solids which are later replaced by foods eaten by older family members. Rice, mungbean, apple and walnut are food materials that are readily available in Jammu and Kashmir and have a promising nutritional attributes. Brown rice, which is hulled directly from rough rice, consists of bran layers (6-7 per cent), an embryo (2-3 per cent) and an endosperm (about 90 per cent). Compared to white rice, brown rice is not only richer in basic nutritional components but also contains more bioactive components, such as ferulic acid, γ-oryzanol and gamma aminobutyric acid (GABA), which exists mainly in the germ and bran layers.

Legumes are principal source for the dietary profile of low-income groups of people in developing countries. Among legumes, mungbean (*Vigna radiata*) is known to have better digestibility and relatively lower antinutritional factors It is also an excellent source of phenolics, flavonoids and other antioxidants besides being rich in dietary fibre, carbohydrates, energy, vitamins and minerals such as vitamin B_6, iron, magnesium, phosphorus, potassium and copper.

Fruits have many health benefits and especially apple (*Malus domestica*) is one of the most widely cultivated tree fruit. Apples have very strong antioxidant activity mainly due to the polyphenols, such as quercetin, catechin, phloridzin, and chlorogenic acid present in them. Apple consumption is also effective in treating infant intestinal disorders, such as diarrhea and dysentery and helps in weight reduction in adults. Walnuts are a good source of high quality protein and contain 18-24 per cent protein on a dry weight basis. Walnuts are therefore, being considered as an additional source of plant protein for use in human food products

Methodology

Seven composite weaning foods were formulated using varying amounts of raw materials by considering nutrient and caloric values of each ingredient in order to meet the nutrient requirement of infants. The brown rice, mungbean, apple pulp and walnut powder were blended separately in different ratios (Table 1) with each other along with 3 g of skim milk powder.

Table 1: Treatment Details for Weaning Food

Treatment	*Brown Rice (Per cent)*	*Mungbean (Per cent)*	*Apple Pulp Powder (Per cent)*	*Walnut Powder (Per cent)*
T_1	100	-	-	-
T_2	85	5	5	5
T_3	80	10	5	5
T_4	75	15	5	5
T_5	70	20	5	5
T_6	65	25	5	5
T_7	60	30	5	5

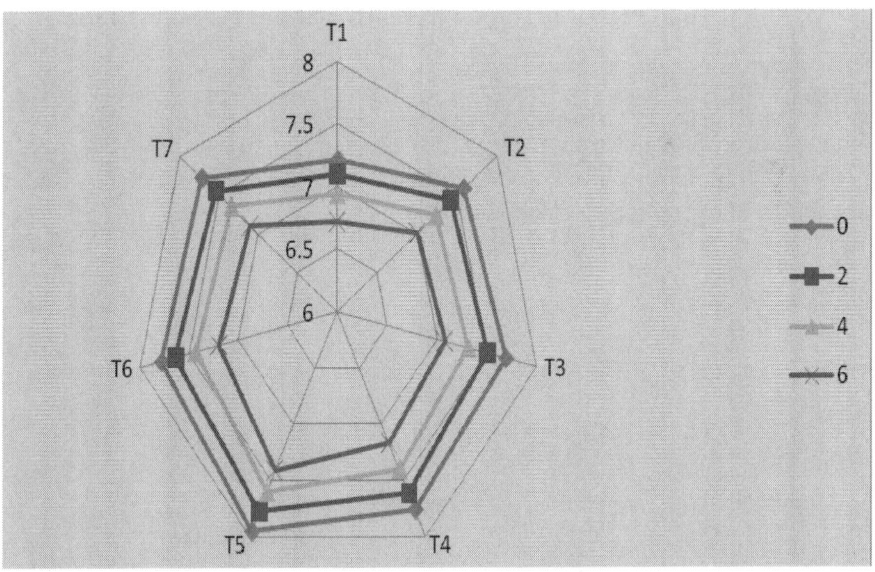

Effect of Treatment and Storage on Overall Acceptability of Weaning Food.

Recommendation

A method has been standardized for developing nutritionally rich weaning food by blending brown rice, mungbean, apple pulp powder, walnut in the different ratios. On basis of sensory evaluation 70:20:5:5 ratio was adjudged best having crude protein (11.57 per cent), crude fat (5.04 per cent), crude fibre (2.26 per cent), iron (4.68 mg/100g) and calcium content (68.96 mg/100g).

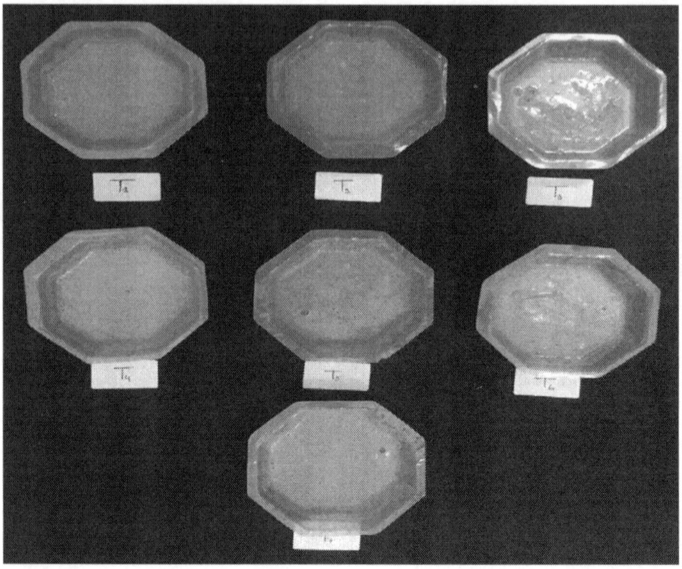

Weaning Food Prepared from different Treatment Combinations.

Benefits

The introduction of supplementation in terms of weaning foods prepared from easily available and low cost ingredients is of vital importance to meet the requirements of the growing children.

Division of Agricultural Engineering

39. Weeding-cum-Earthing Up Equipment

Contact Scientist: **Dr. J.P. Singh**

Objective

Development and evaluation of low cost weeding- cum-earthingup equipment and its comparison with other existing weeding equipment.

Introduction

Weed control is often the most important agricultural task facing farmers in the country. Weeding operation is labour intensive and accounts for about 25 per cent of the total labour requirement which is usually 900-1200 man-hr/ha. The most common methods of weed control are mechanical, chemical, biological and cultural methods. Out of these four methods, mechanical weeding either by hand tools or weeders are most effective. In Indian agriculture, it is a very difficult task to weed out unwanted plants manually as well as using bullock operated equipments which may further lead to damage of main crops. The various types of weeding equipment such as khurpi, push and pull type weeder, dryland weeders, wheel hoe weeder, animal drawn blade hoe weeder and power weeders are used for mechanical weeding or used for burying cutting and uprooting weeds.

Most small and medium scale farmers, which account for nearly 93 per cent of the total number of farmers in India; cannot afford the power weeding through power weeder due to their high initial and operating costs. Since, the size of farm holdings of the J&K farmers is too small to justified the use of big mechanical weeders on their farms. The effective weeding cum earthing up technology can contribute to increase production of small farmers through timely and good quality weeding as well as earting up operation by introducing the improved weeding technology on the farm. There was no efforts have been made on weeding -cum -earthing up technology which can be useful for both weeding and the same time earthing up operation. Therefore, Keeping above points in view a low cost weeding-cum-earthing up equipment was developed for weeding as well as earthing-up operation simultaneously.

Methodology

A low cost weeding- cum-earthing up equipment was developed and fabricated. The developed equipment is consisting of two units *viz.* weeding unit and earthing-up unit. The components of weeding units are such as main frame, handle, weeding blade and ground wheel. The earthing-up unit consisted of two no. of discs to cut and invert the soil. A view of weeding-cum-earthing-up equipment developed in software is shown in Figure 1. The developed equipment (Figure 1) was evaluated and compared with other existing weeding tools/equipments such as and weeding with Khurpi, Wheel hand hoe, Push pull type weeder and Cycle hoe in the field.

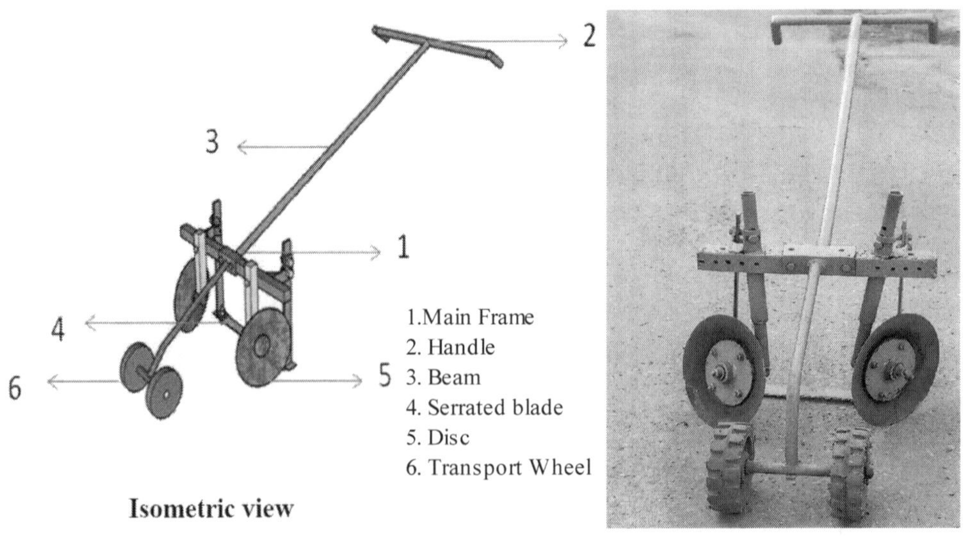

Figure 1: Views of Developed Equipment

Findings

A weeding-cum-earthing-up equipment was developed in the Division of Agricultural Engineering under M.Tech. thesis research work. This equipment consists of two main units *viz.* first weeding unit and second soil cutting and earthing-up unit. A serrated blade is used for cutting/weeding of the weeds and two discs are used for earthing-up operation. It was evaluated on maize crop having sandy loam soil at Advance Centre for Rainfed Agriculture, Dhiansar, SKUAST-Jammu. The weeding efficiency and field capacity of this equipment was found 90.58 per cent and 0.041 ha/h, respectively. It is manual operated small equipment which can be used for weeding-cum- earthling up operation in wide spaced crops.

Impact and Suggestion

The developed equipment can perform both the operations weeding and earthing-up simultaneously, which save labour, time and energy. The developed equipment can be able to reduce the cost of cultivation and increase the income of farmer. Further, research work may be required to reduce the weight of equipment by using lightweight materials. Since the weeder was designed for low cost, the weeder was made manual but it may be made motorized. The developed equipment may be evaluated in other wide spaced crops and type of soil.

40. Development and Evaluation of Automatic Timer Based Variable Speed Device for Sprinkler System

Contact Scientist: **Dr. Sushmita M. Dadhich**

Objectives

1. Development of Automatic Timer Based Variable Speed Device (VSD) for Centrifugal pumps.
2. Field Evaluation of the VSD through Sprinkler System
3. Prepare guidelines on VSD on the basis of infiltration rates under various soils.
4. Cost analysis of the VSD

Introduction

Irrigation pumps usually runs at a constant discharge rate which delivers more water than desired and water runoffs which consumes more time, energy and water. An Automatic Timer Based Variable Speed Device (VSD) has been developed for existing 3 phase centrifugal pumps *i.e.* for 5 hp and 10 hp, respectively. The developed device is compatible with existing electrical irrigation pumps for rotating head sprinkler irrigation system. The developed device allows pump to run with variable discharge rate for shorter periods with different RPMs (frequencies) that minimizes the evaporation losses during sprinkler irrigation system. This will also reduce the operation time, energy consumption and maintenance cost of the pump. The developed device is very useful in agriculture sector. It can also increase the efficiency of the sprinkler irrigation system vis-a-vis saves water, energy and time. The device is an attachment of the existing electrical irrigation pumps used for irrigation in agriculture sector.

The developed device allows pump's motor to run with variable RPMs and thus the pump's discharge according to soil's infiltration rate for shorter period with different RPMs (frequencies) that minimizes the evaporation losses during sprinkler irrigation system. This will also reduce the operation time, energy consumption and maintenance cost of the pump. The different components of the system have been integrated in one box as a unit. The whole system has been integrated first time as a handy unit in one box and is unique in operation and technology for small and portable irrigation pumps. This can be carried anywhere as a suitcase and fitted very easily with the pumps in the field by farmer himself. This is unique and innovative idea to set all the desired component in one compact box and the whole system is available as a box unit which is very handy, compact and easy to use.

Methodologies

The main component of the device is variable frequency controller which allows the motor of the pump to run at different RPMs in desired time slots. Variable frequency controller is connected with different accessories to measure the different variables and compatible with existing electrical irrigation pumps. The All components are connected with MCB. The output of the MCB is connected to two numbers of contactor. One contactor for DOL (Direct Off Line) with overhead relay and second to the input supply contractor of VFD. One contactor is connect to the output of the VFD. The output contactor is used because when system is operated at DOL Starter, in that condition, supply should not be fed to the output of VFD otherwise VFD will Damage.

The different components of the system have been integrated in one box as a unit. In this way, existing irrigation pump can be operated by either one of the system. In the box, there are three indicating lamps for power indicators and there is one selector switch for selection of motor run either at DOL (Direct Off Line) or VF. One number of multifunction energy meter is also mounted on the top door of the box. Three numbers of current transformers are fitted in the incoming supply of the system. The output of the current transformers are connected to the energy as well as incoming current supply for power calculation.

Data/Table

Energy consumption by 5 hp and 10 hp irrigation pumps with VSD are 3.67 per cent and 1.77 per cent less with respect to existing irrigation pumps respectively. 5 hp irrigation pump with VSD was operated for one hour at different frequency, 10 minutes time interval and rated voltage at 350 volt and it can be noticed that energy consumption was less for irrigation pump with VSD *i.e.* 22.98 per cent less energy consumption when irrigation pump is used with VSD as compared to existing irrigation pump. Similarly, 10 hp irrigation pump is operated at various frequencies for 10 minute intervals for one hour and the variation in energy consumption when 10 hp irrigation pump is operated with and without VSD. Energy consumption was 40.43 per cent less when irrigation pump is operated with VSD as compared to existing irrigation pump.

Photos

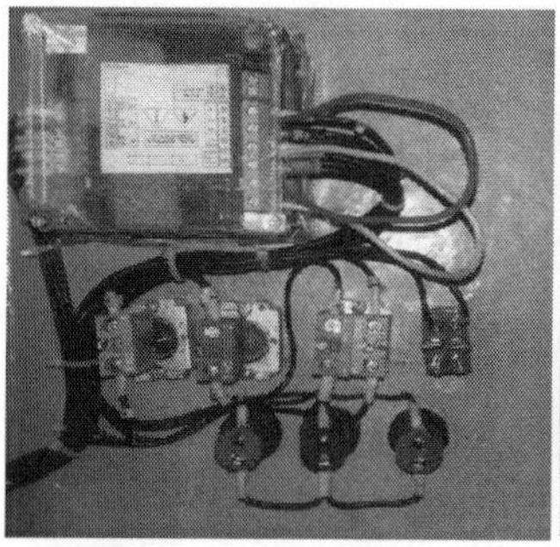  

Developed VFD for Pumps

Recommendations

Private sector companies/manufacturers of electric motors and pumps may start producing/commercializing it as an attachment device with existing pumps for sprinkler irrigation system used in agricultural sector. Initially, the possible interest of large scale farmers and government agricultural farms, R and D institutions of government and private sector working on water and energy saving technologies in agricultural sector. Private sector companies may start producing it as an attachment device with existing pumps for sprinkler irrigation system.

Impact/Benefits

The concept of variable speed device is new in agricultural sector. This technology has very good scope in agriculture water management sector as it can save energy as well as precious water for irrigation. The newly developed device can be operated for both arrangement with existing irrigation pumps (i.e either run at rated RPM or at variable RPM). The developed device is very useful in agriculture sector. It can reduce the maintenance cost of the irrigation pump and increase the efficiency of the sprinkler irrigation system vis-a-vis saves water, energy and time. It fulfils its target rating ₹value for money'.

41. Design and Development of a Tractor Operated Soil Compaction Measurement Device

***Contact Scientist:* Er. Hemant Dadhich**

Objectives

1. Reviews of existing penetrometers for measurement of cone index of soil.
2. Modification in existing penetrometer for automatic measurement of cone index of soil by adding sensors.
3. Evaluation of the developed instrument.

Introduction

A reduction in yield or yield potential is the most significant effect that soil compaction has on crops. The inability of roots to penetrate compacted soil layers will result in decreased yield. It was reported that soybean root growth was decreased by 10 per cent when the soil penetrometer resistances were 0.52 MPa and by 50 per cent when the soil penetrometer resistance were 1.45 MPa in sandy loam soil. With less root penetration into the soil, root mass is reduced and a plant's ability to take up nutrients is reduced. As a result of surface soil layers having lower water storage capacities, plant roots remain closer to the surface and are therefore more susceptible to drought. Compacted soils often have higher subsurface soil moisture contents because soil water is unable to drain away freely and air movement in the soil is restricted. This reduction in internal drainage generally leads to surface ponding which may drown the crop.

Soil compaction is a human-induced problem that can affect the soil-plant-water relationship. It can be prevented through better soil practices that do not disrupt the natural soil architecture. The elevation of soil compaction is best achieved by reducing the number of tillage operations preferably by adopting a zero tillage system, avoiding traffic and tillage when the soil is moist or wet, using the ₹plastic limit test for soil cultivation, practicing controlled traffic farming, sub-soiling for elevation of subsoil compaction, using large diameter narrower tyres, which compact less area than small wide tyres or dual tyres. The curing of problem required identifying the affected areas otherwise, the valuable input will waste by user and soil compaction is not an exception.

Methodologies

To address on specific problem on soil compaction measurement, a soil compaction measurement device using hydraulic system of tractor for pushing penetrometer inside soil at recommended constant speed and at the same time measurement was recorded by data logger with the help of sensors has been developed in the project. The developed machine used the hydraulic power of the tractor attached to the machine and able to measure the no. of data in a day by more accuracy vis-à-vis recording the data for further analysis. Soil samples for soil compaction up to 600 mm can be recorded by the machine.

A two way hydraulic cylinder fitted on a frame in such a way that both the points of the cylinder can be attached with the tractor's hydraulic points for power output. A 5kN ₹S' type load cell is fixed at lower end of the hydraulic cylinder and a 100 cm long penetrometer for insertion into the soil up to 60 cm depth is attached with this load cell. In this way, the entire load exerts on the penetrometer can be read by load cell. A static cone penetrometer with a 300C cone (recommended by the American Society of Agricultural Engineers), as the standard measuring device for characterizing the penetration resistance of soils (ASAEa, 1999) is attached to hydraulic cylinder's lower end through a ₹S' type load cell for measurement of load or resistance to the soil. A resistive displacement sensor is also fixed with the frame and the extension wire of the sensor is fixed at the joint of cylinder and penetrometer, which moves along with the penetrometer at the time of insertion and exit. The load cell and displacement sensor are calibrated for measurement of load and displacement, respectively while probe going into soil up to 600 mm.

A two port data logger is attached with load cell and displacement sensor. The load/resistance (measured by the load cell) corresponding to displacement measured by resistive displacement sensor is recorded in data logger as the probe/cone enters constantly into soil up to 600 mm. Operator of the machine (tractor operator) controls the entering and exiting of the probe into and out of the soil by a direction control valve fitted with the frame near the operator. Operator has to push the start button on data logger and change the direction of direction control valve for insertion of the probe in to the soil. The probe/cone goes in to the soil up to 600 mm and stops and then operator has to change the direction of direction control valve for pulling bank the probe/cone.

The data logger has two lines of display showing Test no. (the number of test recording in a day/series), date and time in top line and displacement in mm and load/resistance in kg force in lower/second line. Total 50 numbers of readings can be recorded at a time in the data logger. After recording the readings, the data from the data logger can be transferred to the computer through data cable. After this, the data can be erased in the data logger and the system is ready for next recordings. All the above system has been developed in coordination with industrial partner, Ms. Siya Instruments, Udaipur, Rajasthan. Now, the machine has been tested successfully and is ready for commercial production through industrial partner.

A soil compaction measurement device using hydraulic system of tractor for pushing penetrometer inside soil at recommended constant speed and at the same time measurement was recorded by data logger with the help of sensors has been developed in the project. The developed machine used the hydraulic power of the tractor attached to the machine and able to measure the no. of data in a day by more accuracy vis-à-vis recording the data for further analysis. Soil samples for soil compaction up to 600 mm can be recorded by the machine. The above system has been developed with industrial partner, Ms. Siya Instruments, Udaipur, Rajasthan. The total cost of the prototype is about Rs. 1,74,200.00 only.

The developed machine was tested at different locations for measurement of the soil compaction. Specifically, for final testing, the machine was tested at four locations *i.e.* Seed production research farm at *Chatha* of the University, Veterinary Science and Animal Husbandry research farm at *RS Pura* of the University, Advanced Centre for Rainfed Agriculture research farm at *Dhiansar* of the University and Advanced centre for Horticulture Research at *Udaiwala* campus of the University. About 50 sets of data were taken at each location and recorded in the data logger. Some of the data sets selected randomly and are presented in annexure 1-4, respectively.

It has been observed that the resistance of the soil is increases when we go deep into the soil from top. Sometimes, it decreases but the major trend is continuously increasing for all the locations. When the data were plotted in the form of graphs, it shows that the soil resistance increases up to maximum depth and when probe comes out from the soil, it decreases in the negative side as the probe encounters soil resistance during withdrawing from the soil. The soil resistance trends were different for all the locations.

The Prototype of the Machine.

The Developed Machine with Attached to Tractor.

Recommendations

A sensor based soil compaction measurement device using hydraulic system of tractor has been developed with industrial partner, Ms. Siya Instruments, Udaipur, Rajasthan with the financial assistance from Department of Science and Technology, GoI. It has been observed that the resistance of the soil is increases when the probe goes deep into the soil from top. Sometimes, it decreases but the major trend is continuously increasing for all the locations. The total cost of the prototype is about Rs. 1,74,200.00 only which may reduce with the orders of the machine and produce in bulk. Further, the machine can be modified/improved for better data logging and measurement at somewhat reduced cost. The frame and hydraulic system of the machine can be designed for universal uses and compatible for all available models of the tractors.

Impact/Benefits

A sensor based soil compaction measurement device using hydraulic system of tractor has been used for the resistance of the soil upto 60 cm depth into the soil. The total cost of the prototype is about Rs. 1,74,200.00 only which may reduce with the orders of the machine and produce in bulk. The machine is very beneficial for researchers, scientists, students, KVKs and other big farmers who want to use soil compaction data for their research and soil profile analysis. The machine gives the variable soil compactness at various depths and this will helpful to plan the crop to be grown in the particular area.

Division of Fruit Science

42. Characterization of Walnut (*Juglans regia* L.) Germplasm in Jammu Province

Contact Scientist: **Dr. Parshant Bakshi**

Introduction

The state of Jammu and Kashmir is the major producer of walnuts in India, in fact, the entire quantity of walnuts the country exports is from J&K state. There are many open pollinated seedlings on their own roots in temperate regions of Jammu province, and these are an important source of genetic diversity. High variability in size of crown, stem, leaves, shape and size of fruit, colour and thickness of shell and kernel has been reported for walnut trees for different regions. Breeders over the years have exploited the variation amongst the existing seedling trees to select superior genotypes with desirable traits. Besides, improved cultivars were introduced from other countries and after evaluation, some recommendations have been made. Therefore, there is a need to study the genetic diversity in walnut in Jammu region.

Objectives

1. To study the genetic diversity in walnut germplasm of seedling origin through morphological, biochemical and molecular characterization.
2. To identify and select the most promising accession of walnut for conservation and registration.

Methodology

The Kishtwar district of Jammu and Kashmir that lies in north-east corner of Jammu region at 33° 19′12″N and 75° 46′12″E coordinates was extensively surveyed to select the promising genotypes of walnut. The hot spots of genetic diversity of walnut were selected and on the basis of preliminary field observations and interactions with owners, the irregular bearing trees with extremely hard shelled nuts were rejected. On the basis of preliminary survey, 200 walnut trees studied and 120 walnut trees were marked and evaluated. Finally, 84 trees were selected at fruit maturity on the basis of size, thickness of shell. The observations on various tree characters *viz.*, foliage, flowering, yield, nut and kernel characters of these trees were recorded as per the walnut descriptor UPOV-TG/125/6 (1999) Geneva. Genetic diversity was studied using SSR molecular markers. A set of 25 random primers were selected for use in amplification of genomic DNA.

The data recorded was statistically analysed with the help of INDOSTAT (version 7.00, Hyderabad, India). Analysis of variance, cluster analysis based on Tocher's method using Squared Euclidean distance was performed using the statistical software Indostat and Statistical Package for Agricultural Research (SPAR) version 7.0 programme. The genetic divergence was calculated according to Mahalanobis D^2 statistics (1936). The data recorded during the investigation was also subjected to the statistical analysis through principal component analysis. The divergence was studied using Kaufman and Rosseau (1990).

Results

The genotypes JWSP-06, JWSG-43 and JWSD-59 are diverse at morphological and molecular level and have good nut attributes, mineral content and fatty acids (linoleic acid) content. The selection JWSG-43 has been found to be more diverse than other two with respect to different nut and kernel characters. The details of physic-chemical characteristics and nuts and kernels of these three selections are given below.

Characteristics	*Walnut Genotype JWSP-06*	*Walnut Genotype JWSG-43*	*Walnut Genotype JWSD-59*
Nut weight (g)	19.18	22.66	17.22
Nut length (mm)	44.32	39.57	43.36
Nut width (mm)	42.17	40.62	39.28
Kernel weight (g)	10.22	14.00	8.89
Percent kernel recovery	53.28	61.78	51.63
Shell thickness	1.44	1.05	1.47
Oil content (per cent)	61.50	61.07	62.43
Protein	22.00	16.40	23.00

WALNUT GENOTYPE- JWSG-43

Name of farmer: Shri Kashmiri Lal **S/o** Shri Chuni Lal **Place**: Galhar, Kishtwar

Latitude: 33° 20.586 **Longitude**: 075°55.770

A

B

C

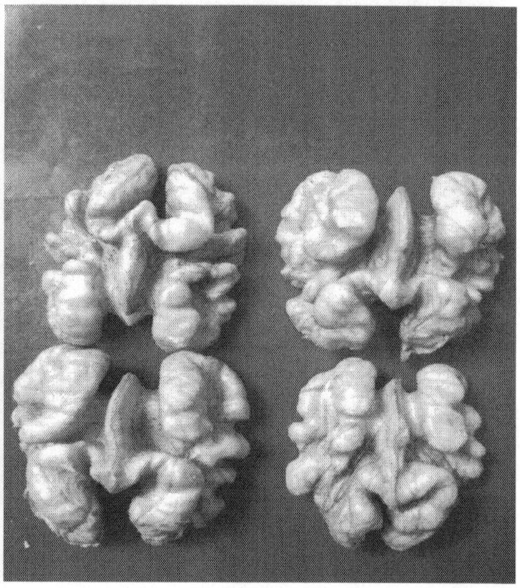

D

Plate 1: A: Mature Tree B: Fruit C: Nut D: Kernel.

WALNUT GENOTYPE- JWSD-59

Name of farmer: Shri Mukhtiyar Ahmad Sheikh **S/o** Shri Wali Sheikh

Place: Lower Janakpur, Dachan **Latitude:** 33° 29.311 **Longitude:** 075°50.477

A

B

C

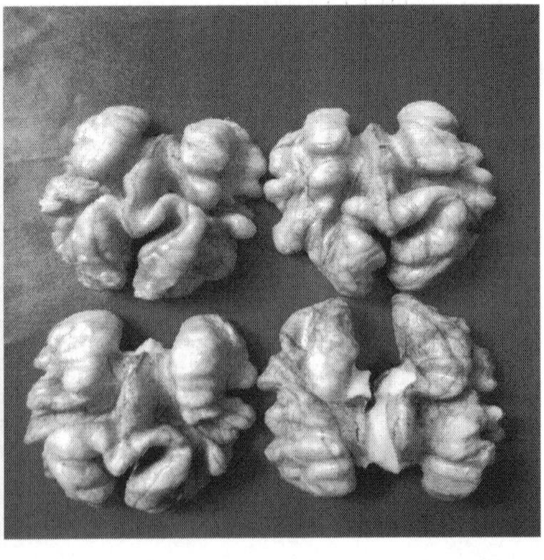

D

Plate 2: A. Mature Tree; B. Fruit; C. Nut; D. Kernel.

WALNUT GENOTYPE- JWSP-06

Name of farmer: Sh. Gasaon Ram **S/o** Sh. Devi Saran **Place**: Gow, Padder

Latitude: 33° 12.0.862 **Longitude**: 76°13 140

A

B

C

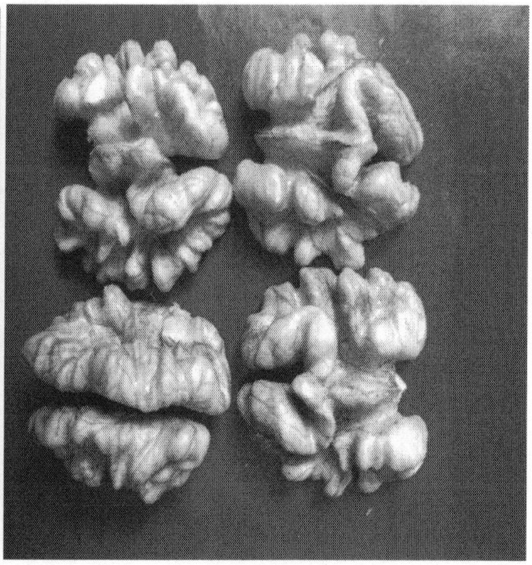

D

Plate 3: A. Mature Tree; B. Fruit; C. Nut; D. Kernel

Recommendations

The genotypes JWSP-06, JWSG-43 and JWSD-59 are diverse at morphological and molecular level and have good nut attributes like size, kernel recovery percentage, market acceptability, oil content percentage, mineral content and fatty acid (linoleic acid) content. The selection JWSG-43 has been found to be more diverse than other two with respect to different nut and kernel characters. It is further recommended that these promising seedling strains are having the potential for quality walnut production and may be used for further multiplication and breeding programmes in order to enhance the production and quality of walnut in the Jammu province of J&K state.

Impact/Benefits

These promising seedling strains are having the potential for quality walnut production and can be used for further multiplication to enhance the production and quality of walnuts in the Jammu province of J&K state. The bud wood of JWSP-06 had been collected and grafted at Nut Centre of SKUAST-Jammu at Rajouri.

43. Characterization of Pecan Nut (*Carya illinoensis* K.) Germplasm of Jammu Province

***Contact Scientist:* Dr. Arti Sharma**

Objectives

1. To study the genetic diversity in pecan nut germplasm of seedling origin in Jammu province through morphological, biochemical and molecular characterization.
2. To identify and select the most promising accession of pecan nut for conservation and registration.

Introduction

The Pecan nut [*Carya illinoensis* (Wangenheim) C. Koch] belongs to the family "Juglandaceae". In India, it is mainly grown in Jammu and Kashmir and Himachal Pradesh. The total area under pecan nut production is increasing due its high economic returns and adaptation to intermediate zone (914-1829 m). In India, it is grown over an area of one thousand ha with the production of about 1 thousand MT. The area under pecan nut in Jammu region is around 414 ha with production of about 86 MT in 2015-16. Pecan nut has the recognition of being called "queen of nuts" because of its excellent kernel characteristics. Pecan having (sweet flavour, 65-70 per cent fats, 8-10 per cent proteins, high in phosphorous, potassium and vitamins A, C, E and B complex) is superior to walnut in quality and thrives best in areas which are considered lower and hotter for walnut cultivation. The present study was undertaken with an aim of surveying different areas of Jammu, Rajouri and Poonch districts for selecting pecan nut genotypes and to estimate the extent of genetic relationship among them by morphological, biochemical and molecular characterization.

Methodology

The present investigation was carried out during the year 2015-2016 at different selected pecan nut growing areas of Jammu province. Survey was done from major pecan nut growing districts *viz.,* Jammu, Rajouri and Poonch of Jammu province of Jammu and Kashmir state purposely during 2015 and 2016 to select promising accession among the diverse pecan nut (seedling) genotypes and assess variability in their physiological and morphological characteristics. During the survey to get first hand information local inhabitants and district horticulture offices were consulted about production and availability of pecan nuts of seedling origin in areas at various places. The location was selected with respect to the availability of diversity in seedling pecan nut genotypes. On the basis of preliminary field observations and interactions with owners, the trees which were bearing extremely small sized, poor quality nuts and low yielding were not included. Finally, plants of sixty (60) seedling origin pecan nut genotypes with divergent characters were selected at fruit maturity stage on the basis of size and kernel quality. Codes were allotted to each selection on the basis of their location and geo tagging was done on selected plants. Regular visits were made during the period of flowering, fruit setting, fruit maturity and ripening stage. The observations on tree characters, leaf characters, inflorescence characters, nut characters and kernel characters were recorded by using the pecan nut descriptors given by UPOV, 2004 for selected seedling trees.

Biochemical characterization was done at Quality control and Quality assurance laboratory of IIIM (Indian Institute of Integrative Medicine) and molecular studies for diversity analysis of selected pecan nut selections was done at molecular Lab at Division of Fruit Science, in Sher-e- Kashmir University of Agricultural Sciences and Technology of Jammu.

Results

Sixty seedling trees of pecan nut selected during survey on the basis of variability for various morphological and phenological traits constituted the material of study as per the descriptor given by UPOV. The selections were analysed for various statistical parameters such as mean, range, coefficients of variation, heritability, genetic advance, correlation and genetic diversity using multivariate statistical tools.

The pecan nut selections evaluated during investigation showed wide variability for various descriptive and quantitative characteristics. Tree vigour varied as strong, medium and weak. Nut shape in ventral view varied from circular, elliptic, oblong, obovate to ovate and kernel colour varied from light, medium to dark. Highest heritability (94.8 per cent) was recorded for kernel length and lowest (43.1 per cent) for shell thickness. Maximum genetic gain (46.61) per cent) was observed for kernel length and lowest (16.85 per cent) was recorded in case of thickness of shell.

The correlation coefficient data revealed that lateral nut width has highest and significant genotypic (0.992) and phenotypic (0.676) positive correlation with nut weight. Path analysis revealed that lateral nut width has maximum (0.517) direct effect on nut weight and kernel weight has maximum (0.404) indirect effect on nut weight via lateral nut width. Divergence analysis of sixty pecan nut selections, using Mahalanobis D2 statistics grouped selections into eight clusters. Cluster VII followed by cluster VIII and III recorded maximum mean values for all nut and kernel characteristics. Maximum (20.36) intra cluster distance was observed for cluster VI, which reveals that maximum diversity is found within selections of cluster VI and maximum (266.26) inter cluster distance is between cluster VII and V which shows that selections of these clusters are most diverse to each other.

Biochemical analysis was done for selections falling in clusters VII, VIII and III. Maximum content of Calcium (74.80 mg/100g), Zinc (5.93 mg/100g), Copper (15.83 mg/100g), Vitamin B1 (0.67 mg/100g), protein (11.56 per cent), oleic acid (76.07 per cent), stearic acid (3.16 per cent) and oil content (74.78 per cent) was observed in selection SKJPM21. Magnesium, Vitamin B6 (0.25 mg/100g), Vitamin B9 (26.16 mcg/100g) and Linoleic acid (41.52 per cent) was recorded maximum in selection SKJPP8. While Manganese (9.47 mg/100g), total phenols (110.25 mg CAE/g, condensed tannins (60.36 mg CE/g), antioxidant capacity (ORAC assay) (828.00 μmol TE/g) and antioxidant activity (DPPH assay) (140.23 mg TE/g) and was observed maximum in selection SKJPP25.

For molecular characterization sixty selections were subjected to DNA profiling using 20 SSR markers. Out of 20 SSR markers only 19 showed polymorphism. Amplified DNA fragments ranged from 50-280 base pairs and dissimilarity coefficient ranged from 0.08 to 0.89. Maximum dissimilarity coefficient value (0.89) between selections SKJPP8 and SKJPP25 and SKJPM21 and SKJPP25 reveals that these selections are genetically diverse and can be used for further breeding programme. UPGMA based dengrogram was constructed based on SSR analysis which grouped sixty selections into 3 major clusters.

Recommendations

Based on the findings of the Morphological and biochemical investigations, it can be recommended that these three seedling pecan nut selections SKJPP8 from Shankar nagar, Poonch, SKJPM21 from Miran Sahib, Jammu and SKJPP25 from Chandak, Poonch were best for various tree, leaf, inflorescence, nut and kernel characters. On the basis of comparison studies between seedling pecan nut selections (SKJPP8, SKJPM21 and SKJPP25) and commercial pecan nut varieties (Mahan, Schley and Burkett) in which morphological and biochemical studies revealed that seedling pecan nut selections are better than commercial cultivars of pecan nut. Molecular studies established that these three pecan nut seedling selections are genetically diverse and do not belong to any of the commercial varieties. So, these three pecan nut selections with excellent traits of interest and high genetic dissimilarity can be used directly as cultivar after completing all codal formalities for variety release or can be further used in breeding programmes to get segregates.

Impact

These three best seedling pecan nut selections *viz.* SKJPP8, SKJPM21 and SKJPP25 having broad genetic base can serve as pertinent genetic sources for initiating trait-specific breeding programmes and have greater potential for commercialization of pecan nut in Jammu Division.

Division of Agroforestry

44. Standardize Harvesting Time, Pre-sowing Seed Treatment, Identification of Superior Genotype of Harad and its Vegetative Propagation

Contact Scientist: **Dr. K.K. Sood**

Methodology

Harad (*Terminalia chebula*) is multipurpose and medicinal tree species, the fruit of which is used in a number of Ayurvedic medicines to treat common ailments such as heart burn, flatulence, dyspepsia, liver and spleen disorders, asthma and constipation *etc.* The fruits are rich in tannin and are also used for tanning in leather industry. Large fruit size fetches higher price in the market. Besides fruit, it is commonly used as timber, leaf fodder, fuelwood and also an important bee-flora.

Interest in cultivation of this species is constrained by:

- Poor regeneration,
- Long gestation period,
- Higher fruit set heights,
- Unavailability of superior genotypes with largest fruit size
- Lack of standardised agro-technology.

The present study investigates the effect of different harvesting time and seed treatments on seed germination, identification of superior genotypes and effect of methods of grafting on sprouting success and growth of the seedlings.

Table 1: Effect of Seed Harvesting Time on Seed Germination

Date of Collection	*Mean Germination (per cent)*
November 15th	0 (*0)[a]
December 1st	0 (0)[a]
Dec 16th	0 (0)[a]
Jan 1st	0 (0)[a]
Jan 16th	7.6 (15.9)[b]
Feb 1st	10.0 (18.0)[c]
Feb 16th	10.7 (18.9)[c]
March 3rd	10.3 (18.6)[c]
March 18th	10.3 (18.6)[c]
CD0.05=2 9 7	

*Figures in brackets are transformed values.

Means with different letters indicate $p< 0.05$ Tukey's honestly significant difference for transformed values.

Table 2: Effect of different Seed Treatments on Seed Germination

Treatment	*Mean Germination (per cent)*
Control (No treatment)	9.5 (*17.91)[d]
Concentrated Sulphuric acid dip (minutes)	
5	12.4 (20.4)[d]
10	12.0 (20.2)[d]
15	15.6 (23.2)[d]
20	17.3 (24.3)[c]
25	20.3 (26.6)[cd]
30	17.8 (24.8)[cd]
Cow dung scarification (days)	
15	25.4 (30.1)[c]
30	38.7 (38.4)[b]
45	37.7 (37.9)[b]
Seed coat breaking	73.6 (59.2)[a]
CD0.05 =5.72	

*Figures in brackets are transformed values Means with different letters indicate $p< 0.05$ Tukey's honestly significant difference for transformed values.

Table 3: Variability in Fruit Characteristics for Identification of Superior Genotype

CD	*0.05*	*3.82*	*5.41*	*0.64*	*0.42*	*3.19*	*5.25*	*11.05*
Province	*Location of Mother Tree*	*Fruit Weight (g)*		*Fruit Size (cm)*		*Pulp Weight (g)*		*Number of Fresh Fruits per kg*
		Fresh	*Dry*	*Length*	*Dia-meter*	*Fresh*	*Dry*	
Jammu and Kashmir	Raya	15.59^{e}	8.10^{c}	3.72^{c}	3.29^{ab}	12.20^{f}	6.53^{a}	71^{a}
	Khadargal	20.16^{d}	7.65^{c}	5.59^{b}	2.66^{d}	17.22^{d}	5.54^{ab}	56^{b}
	Miran-Sahab	20.74^{d}	9.41^{c}	5.27^{b}	2.87^{b}	18.37^{cd}	7.97^{a}	51^{bc}
	Mangloor	18.34^{d}	7.05^{c}	3.38^{c}	2.24^{d}	15.06^{e}	5.16^{b}	55^{b}
	Keri	18.04^{d}	8.50^{c}	3.43^{c}	2.65^{cd}	14.27^{e}	6.77^{a}	56^{b}
	Dhoon	18.86^{d}	6.93^{c}	3.62^{c}	2.47^{d}	15.50^{de}	4.79^{b}	53^{b}
	Mathwar	44.60^{a}	21.75^{a}	7.90^{a}	3.38^{a}	34.78^{a}	10.79^{a}	24^{e}
	Rabta	35.80^{b}	18.00^{b}	5.67^{b}	3.55^{a}	27.21^{b}	8.71^{a}	29^{de}
	Sandhi	26.46^{c}	13.36^{b}	5.69^{b}	3.00^{bc}	19.06^{c}	6.29^{a}	40^{cd}
	Sarore	27.00^{c}	14.85^{b}	5.49^{b}	3.03^{b}	21.33^{c}	6.39^{a}	39^{d}
	Nandini	28.00^{c}	14.28^{b}	5.40^{b}	3.06^{b}	21.34^{c}	6.42^{a}	37^{d}
	Joginder Nagar	19.91^{d}	8.93^{c}	5.08^{b}	2.66^{c}	15.81^{d}	6.91^{a}	52^{b}
Himachal Pradesh	Dhualakuan	19.42^{d}	6.21^{c}	3.56^{c}	2.24^{d}	16.19^{d}	4.42^{b}	52^{b}
	Panchrukhi-Palampur	18.86^{d}	5.94^{c}	3.59^{c}	2.31^{d}	15.98^{d}	4.45^{b}	53^{b}

Table 4: Effect of Rootstock Size and Grafting Methods on Growth of above Ground Parameters of Elite Grafts

Grafting Method	*Rootstock Size*		
	Small	*Large*	*Mean*
Patch budding			
FSW	31.16^{aA}	36.31^{aA}	33.73^{A}
SL	63.75^{aA}	82.20^{bA}	72.98^{A}
FLW	46.23^{aA}	55.86^{bA}	19.93^{A}
LL	14.33^{aA}	15.55^{aA}	14.94^{A}
LW	8.75^{aA}	08.55^{aA}	08.63^{A}
NL	20.50^{aA}	24.00^{bA}	22.25^{A}
DSW	12.55^{aA}	15.38^{bA}	13.97^{A}
Cleft grafting			
FSW	31.86^{aA}	37.25^{aA}	34.75^{B}

Grafting Method	*Rootstock Size*		
	Small	*Large*	*Mean*
SL	65.91 ^{aB}	83.75 ^{bB}	74.83 ^{B}
FLW	51.20 ^{aB}	55.96 ^{bB}	21.22 ^{B}
LL	14.05 ^{aB}	15.45 ^{aB}	14.75 ^{A}
LW	9.10 ^{aB}	08.51 ^{aB}	08.83 ^{A}
NL	22.00 ^{aB}	26.00 ^{bB}	24.00 ^{B}
DSW	12.76 ^{aB}	18.42 ^{bB}	15.60 ^{B}
Mean			
FSW	31.51 ^{a}	36.97 ^{b}	33.74
SL	64.82 ^{a}	82.97 ^{b}	73.91
FLW	48.72 ^{a}	55.91 ^{b}	20.58
LL	14.19 ^{a}	15.50 ^{b}	14.85
LW	08.93 ^{a}	08.53 ^{b}	8.73
NL	21.25 ^{a}	25.00 ^{b}	23.13
DSW	12.66 ^{a}	16.90 ^{b}	14.79
	CD0.05		
	Size (S)	*Method (M)*	*S x M*
FSW	1.22	1.22	4.43
SL	3.13	3.13	2.58
FLW	1.24	1.24	0.24
LL	0.17	NS	0.22
LW	0.16	NS	1.71
NL	1.21	1.21	4.16
DSW	2.94	2.94	0.83

Lower case letters indicate results for means comparison between small and large sized root stock for respective characteristic.

Upper case letters indicate results for means comparison between grafting methods for respective characteristic.

Different letters for either comparison indicate $p<0.05$ Tukey's significant difference for the respective characteristics.

Abbreviations: FSW-Fresh Shoot weight (g), SL-Shoot length (cm), FLW-Fresh leaf weight (g), LL-Leaf length (cm), LW-Leaf width (cm), NL- Number of leaves, DSW dry shoot weight (g)

Table 5: Effect of Rootstock Size and Grafting Methods on Growth of Root Characteristics of Elite Grafts

Grafting Method	*Rootstock Size*		
	Small	*Large*	*Mean*
Patch budding			
FRW(P)	29.27^{aA}	36.76^{bA}	33.01^{A}
FRW(S)	09.44 ^{aA}	11.88 ^{bA}	10.66 ^{A}
NSR	46.50 ^{aA}	65.50 ^{bA}	56.00 ^{A}
RL(P)	40.90 ^{aA}	51.60 ^{bA}	46.25 ^{A}
DRW(P)	12.23 ^{aA}	16.08 ^{bA}	14.16 ^{A}
DRW(S)	2.51 ^{aA}	4.17 ^{bA}	3.34 ^{A}
Cleft grafting			
FRW(P)	31.31 ^{aB}	37.24 ^{bB}	34.28^{B}
FRW(S)	12.19 ^{aB}	12.40 ^{bB}	12.29 ^{B}
NSR	49.00 ^{aB}	68.00 ^{bB}	58.50 ^{B}
RL(P)	50.20 ^{aB}	53.75 ^{bB}	51.98 ^{B}
DRW(P)	15.24 ^{aB}	16.27 ^{bB}	15.76 ^{B}
DRW(S)	4.51 ^{aB}	4.53 ^{bB}	4.52 ^{B}
Mean			
FRW(P)	30.29 ^{a}	37.00 ^{b}	33.65
FRW(S)	10.82 ^{a}	12.14 ^{b}	11.48
NSR	47.75 ^{a}	66.75 ^{b}	57.25
RL(P)	45.55 ^{a}	52.67 ^{b}	49.12
DRW(P)	13.73 ^{a}	16.18 ^{b}	14.96
DRW(S)	3.51 ^{a}	4.35 ^{b}	3.93
	Size (S)	*Method (M)*	*S x M*
FRW(P)	0.89	0.89	1.26
FRW(S)	0.52	0.52	0.73
NSR	6.42	6.42	4.35
RL(P)	1.54	1.54	2.18
DRW(P)	0.59	0.59	0.83
DRW(S)	1.32	1.32	1.87

Abbreviations: FRW(P) - Fresh weight of primary roots (g), FRW(S)- Fresh weight of secondary roots (g), NSR-Number of secondary roots, RL(P)- Length of primary root (cm), DRW(P)- Dry weight of primary roots (g), DRW(S)-Dry root weight of secondary roots (g)

Lower case letters indicate results for means comparison between small and large sized root stock for respective characteristic.

Upper case letters indicate results for means comparison between grafting methods for respective characteristic.

Different letters for either comparison indicate $p<0.05$ Tukey's significant difference for the respective characteristics.

Mother Tree of Raj Harad at Village Mathwar.

Cleft Grafting with Scion from Raj Harad

Raj Harad Plants in Fruiting Stage at University Farm, Chatha.

Recommendation

- The fruits shall be harvested from January to March for better germination of the seeds
- Pre-treatment of the fruits either by extraction of kernels through mechanical means and cowdung scarification for 30 days is advised to enhance germination under nursery conditions.

- "Raj Harad"- a candidate plus tree marked in Mathwar, Tehsil Akhnoor was found superior than other in terms of fruit size and pulp content.
- Cleft grafting on 2 years old scion of harad is advised during the month of April to ensure better success (70 per cent)

Benefit/Impact

- Fruit bearing period has been reduced from 25 years to 06 years.
- The fruit size is larger which fetches comparatively high price in the market.
- Farmers have established harad orchards on their farm in their villages like Ranjan, Jandial Thathi, Kathad (Manwal), Mathwar, Shwa *etc.* and getting remunerative prices in the bearing stage itself.

45. Standardization of Agro-technology of Medicinal Plants: *Aloe barbadensis, Gloriosa superba, Rauvolfia serpentina* and *Silybum marianum*

Contact Scientist: **Dr. L.M. Gupta**

Methodology

Medicinal plants have a great an importance in farm diversification as they are moving from fringe to mainstream use in healthcare and disease remedies being free from side effects. Plant species like Ashwagandha, Sarpagandha, Kalmegh, Senna, Makoi, Giloe, Kalihari, Shatavar, Guggal, Babool, Khejri, Silybum, Amla, Harad, Bahera and Aloe *etc.* are suitable for diversification of farming.

Aloe (*Aloe barbadensis*) due to its ability to avoid water loss by closing stomata survives long periods of drought. The species belongs to family Liliaceae and is commonly known as Kuwargandhal in Punjabi and Dogri, Gheekanwar in Hindi; Ghrita Kumari, Kumari, Ghrit Kumarika in Sanskrit and Indian *Aloe* in English. *Aloe barbadensis* is in high demand owing to its therapeutic, cosmetic and neutraceutical properties. The dried leaf juice is laxative due to the presence of aloin, which is a mixture of glucosides. The aloe gel (mucilaginous pulp of leaf parenchyma) is widely used in cosmetic industry as a skin tonic. The gel also contains over 75 nutrients and 200 active compounds including 20 minerals, 18 amino acids and 12 vitamins.

Table 1: Mean Performance of Growth Parameters in different Accessions of *Aloe barbadensis*

Accessions	*Plant Spread (cm)*	*No. of Suckers Plant^{-1}*	*Leaf Length (cm)*	*Leaf Width (cm)*	*Leaf Thickness (cm)*	*No. of Leaves Plant^{-1}*
IC 111267	30.50	6.00	31.56	3.91	1.17	8.00
IC 111269	36.17	3.01	39.09	4.12	1.20	9.80
IC 111271	75.83	2.00	54.91	7.37	1.31	14.00
IC 111272	59.50	2.67	48.72	4.37	1.28	12.00
IC 111279	47.67	4.44	42.00	4.52	1.18	9.00
IC 111280	38.17	3.67	34.17	4.97	1.19	11.00
IC 112534	69.50	2.33	52.58	7.20	1.30	13.20
IC 471882	45.83	4.00	39.20	5.80	1.20	10.60
IC 471883	55.17	2.89	42.67	5.70	1.22	11.20
IC 471884	51.00	5.45	37.17	5.02	1.25	8.40
IC 471885	42.33	4.78	32.89	5.76	1.18	9.60
$CD_{0.05}$	**8.53**	**1.10**	**6.96**	**1.35**	**NS**	**3.36**

Table 2: Mean Performance of Yield Parameters in different Accessions of *Aloe barbadensis*

Accessions	*No. of Harvestable Leaves Plant^{-1}*	*Fresh Weight Leaf^{-1} (g)*	*Latex Leaf^{-1} (g)*		*Latex Plant^{-1} (g)*		*Gel Yield Leaf^{-1} (g)*	*Gel Yield Plant^{-1} (kg)*	*Leaf : Gel Ratio*	*Leaf Yield plant^{-1} (kg)*
			Wet	*Dry*	*Wet*	*Dry*				
IC 111267	3.33	69.91	0.19	0.09	0.65	0.35	35.36	0.12	1.98	0.23
IC 111269	4.00	100.97	0.31	0.16	1.27	0.66	51.94	0.20	2.07	0.41
IC 111271	6.67	150.00	0.40	0.22	2.66	1.50	83.50	0.56	1.70	0.96
IC 111272	6.67	118.96	0.32	0.18	1.92	1.08	69.33	0.46	1.85	0.80
IC 111279	4.33	97.50	0.32	0.17	1.40	0.77	48.75	0.21	1.96	0.41
IC 111280	5.33	72.91	0.30	0.16	1.59	0.88	40.02	0.21	1.90	0.39
IC 112534	6.00	144.79	0.39	0.20	2.60	1.36	79.28	0.51	1.74	0.90
IC 471882	5.00	106.36	0.20	0.11	1.00	0.57	55.66	0.28	1.89	0.53
IC 471883	6.33	108.15	0.27	0.13	1.71	0.82	63.40	0.40	1.80	0.68
IC 471884	3.67	95.94	0.24	0.14	0.90	0.52	48.03	0.18	2.02	0.36
IC 471885	4.67	76.98	0.22	0.12	1.03	0.56	42.25	0.20	1.78	0.36
$CD_{0.05}$	**1.53**	**16.53**	**NS**	**NS**	**0.57**	**0.21**	**7.64**	**0.14**	**0.13**	**0.27**

Table 3: Effect of Manures and Fertilizers on Growth Parameters of *Aloe barbadensis*

Treatment	*Plant Spread (cm)*	*No. of Suckers Plant^{-1}*	*Leaf Length (cm)*	*Leaf Width (cm)*	*Leaf Thickness (cm)*	*No. of Leaves Plant^{-1}*
T_1 (No treatment)	38.51	1.87	36.68	4.21	1.16	9.11
T_2 [150 kg N (Recommended dose)]	55.17	3.98	50.67	6.47	1.24	12.89
T_3[30t FYM/ha + *Azotobacter*]	43.26	5.75	43.26	5.43	1.20	11.44
T_4[7.5t VC/ha+ *Azotobacter*]	48.20	5.11	47.92	6.06	1.20	12.67
T_5 [75kg N + 15t FYM/ha+ *Azotobacter*]	70.34	3.19	58.91	7.06	1.26	14.55
T_6[75kg N+3.75t VC/ha+ *Azotobacter*]	74.26	2.78	60.27	7.14	1.28	15.33
T_7[50kg N +20t FYM/ha+ *Azotobacter*]	59.50	4.44	52.94	6.53	1.20	13.22
T_8[50kg N+20t FYM/ha]	57.67	4.68	51.05	6.47	1.23	13.00
T_9[50kgN+5tVC/ ha+ *Azotobacter*]	67.11	3.23	56.25	6.67	1.25	14.33
T_{10}[50kg N +5t VC/ha]	63.28	3.67	54.22	6.62	1.24	13.89
$CD_{0.05}$	**8.33**	**0.84**	**6.08**	**0.53**	**N.S.**	**1.97**

Rauvolfia serpentina (Sarpgandha) and *Gloriosa superba* (Agnishikha, Kalihari) are endangered/threatened species found sparsely in tropical and subtropical parts of India. Both the species have got medicinal importance and are widely used in traditional and modern system of medicines. Seed and tubers of *Gloriosa superba* (kalihari) contain active content colchicines, which is used for treating gout and rheumatism besides capable of inducing polyploidy in plants. It is also used for treating cholera, typhus, brigat's disease, piles, leprosy, gonorrhea and chronic ulcers. Roots/tubers of both species have been indiscriminately harvested from their natural habitats without any concerted effort to regrow these plants. Low germination of seeds coupled with indiscriminate harvesting has led to the endangered/threatened status of these species. Ministry of commerce, Govt. of India has put ban on export of raw material of these species due to their threatened/ endangered status, however, the roots of cultivated origin can be sold if their source is certified by the designated authority.

Table 4: Effect of Manures and Fertilizers on Yield Parameters of *Aloe barbadensis*

Treatment	*No. of Harvestable Leaves Plant^{-1}*	*Fresh Weight Leaf^{-1} (g)*	*Latex Leaf^{-1} (g)*		*Latex Plant^{-1} (g)*		*Gel Yield Leaf^{-1} (g)*	*Gel Yield Plant^{-1} (kg)*	*Leaf : Gel Ratio*	*Leaf Yield Plant^{-1} (kg)*
			Wet	*Dry*	*Wet*	*Dry*				
T_1 (No treatment)	4.95	108.27	0.22	0.11	1.10	0.52	50.80	0.26	1.96	0.56
T_2 [150 kg N (Recommended dose)]	6.00	145.96	0.29	0.15	1.80	0.96	74.16	0.52	1.71	0.87
T_3[30t FYM/ha +*Azotobacter*]	5.45	125.63	0.27	0.14	1.64	0.84	59.09	0.37	1.83	0.69
T_4[7.5t VC/ ha+*Azotobacter*]	5.88	136.15	0.29	0.15	1.74	0.92	66.71	0.46	1.75	0.75
T_5 [75kg N + 15t FYM/ ha+*Azotobacter*]	6.79	166.36	0.37	0.18	2.78	1.38	85.56	0.69	1.55	1.08
T_6[75kg N+3.75tVC/ha+ *Azotobacter*]	7.10	171.45	0.38	0.19	2.99	1.52	94.31	0.77	1.51	1.13
T_7[50kg N +20t FYM/ ha+ *Azotobacter*]	6.21	152.69	0.32	0.16	2.15	1.07	76.35	0.56	1.66	0.91
T_8[50kg N+20t FYM/ ha]	6.09	147.70	0.31	0.15	2.10	1.03	75.23	0.54	1.69	0.89
T_9[50kgN+5tVC/ha+ *Azotobacter*]	6.44	159.33	0.34	0.17	2.50	1.26	81.67	0.63	1.59	0.95
T_{10}[50kg N +5tVC/ha]	6.37	156.79	0.34	0.17	2.29	1.19	79.20	0.60	1.60	0.94
$CD_{0.05}$	**1.01**	**10.85**	**NS**	**NS**	**0.50**	**0.39**	**8.34**	**0.10**	**0.15**	**0.12**

Table 5: Effect of Irrigation Scheduling on Growth Parameters of *Aloe barbadensis*

Treatment	*Plant Spread (cm)*	*No. of Suckers Plant^{-1}*	*Leaf Length (cm)*	*Leaf Width (cm)*	*Leaf Thickness (cm)*	*No. of Leaves Plant^{-1}*
T_1 (Rainfed)	43.50	2.00	39.67	4.97	1.12	9.78
T_2(IW_{30mm}/CPE=0.6)	67.11	5.44	47.05	5.76	1.13	12.22
T_3(IW_{30mm}/CPE=0.8)	70.34	5.89	49.42	5.80	1.17	13.11
T_4(IW_{30mm}/CPE=1)	60.28	3.11	53.86	6.54	1.19	10.67
T_5(IW_{40mm}/CPE=0.6)	54.16	6.11	50.00	6.13	1.15	11.44
T_6(IW_{40mm}/CPE=0.8)	62.44	4.11	56.24	6.97	1.20	12.78
T_7(IW_{40mm}/CPE=1)	73.22	3.89	60.25	7.80	1.24	14.22
$CD_{0.05}$	**8.96**	**1.30**	**5.72**	**1.14**	**NS**	**1.70**

Table 6: Effect of Irrigation Scheduling on Yield Parameters of *Aloe barbadensis*

Treatment	*No. of Harvestable Leaves Plant^{-1}*	*Fresh Weight Leaf^{-1} (g)*	*Latex Leaf^{-1} (g)*		*Latex Plant^{-1} (g)*		*Gel Yield Leaf^{-1} (g)*	*Gel Yield Plant^{-1} (kg)*	*Leaf : Gel Ratio*	*Leaf Yield Plant^{-1} (kg)*
			Wet	*Dry*	*Wet*	*Dry*				
T_1(Rainfed)	5.18	110.96	0.20	0.10	1.06	0.54	58.81	0.32	1.94	0.59
T_2(IW_{30mm}/CPE=0.6)	6.22	122.30	0.34	0.15	2.11	0.93	67.27	0.41	1.86	0.74
T_3(IW_{30mm}/CPE=0.8)	6.33	135.82	0.35	0.15	2.23	0.96	76.06	0.49	1.79	0.90
T_4(IW_{30mm}/CPE=1)	5.52	148.03	0.40	0.19	2.21	1.05	81.42	0.45	1.82	0.82
T_5(IW_{40mm}/CPE=0.6)	5.93	140.60	0.39	0.17	2.30	1.00	78.74	0.47	1.80	0.85
T_6(IW_{40mm}/CPE=0.8)	6.71	165.10	0.42	0.20	2.86	1.36	94.11	0.66	1.75	1.09
T_7(IW_{40mm}/CPE=1)	7.01	170.79	0.46	0.22	3.18	1.50	99.06	0.71	1.69	1.21
$CD_{0.05}$	**1.10**	**13.45**	**NS**	**NS**	**0.42**	**0.23**	**9.54**	**0.17**	**0.11**	**0.22**

Table 7: Effect of Fertilizers and Manures on Growth Parameters of *Gloriosa superba*

Treatment	*Days taken to Sprouting*	*Days taken to Initiation of Flowering*	*Plant Height (cm)*	*No. of Leaves per Plant*	*No. of Branches per Plant*
No treatment (T_1)	47.20	81.33	94.75	97.56	1.61
NPK @ 120:50:75 (T_2)	45.91	74.19	120.18	123.30	2.16
FYM @ 20 t/ha (T_3)	43.11	72.59	118.75	138.28	2.12
VC @ 6t/ha (T_4)	38.20	71.52	130.36	154.37	2.20
FYM @ 10 t/ha + 1/2 T_2 (T_5)	39.26	70.06	126.25	128.70	2.25
VC @ 3t/ha + 1/2 T_2 (T_6)	42.83	64.70	136.25	164.39	2.33
FYM @ 13.33 t/ha + 1/3 T_2 (T_7)	39.16	68.25	123.58	166.40	3.41
VC @ 4t/ha + 1/3 T_2 (T_8)	36.66	61.67	144.96	172.03	4.35
$CD_{0.05}$	**5.482**	**8.180**	**9.574**	**10.476**	**0.379**

Table 8: Effect of Fertilizers and Manures on Yield and Yield Attributing Traits of *Gloriosa superba*

Treatment	*No. of Flowers per Plant*	*No. of Fruits per Plant*	*Fruit Length (cm)*	*Fruit Diameter (cm)*	*No. of Seed per Fruit*	*100-Seed Weight (g)*	*Seed Yield per Fruit (g)*	*Seed Yield per Plant (g)*	*Estimated Seed Yield (kg/ha)*
No treatment (T_1)	13.83	6.63	3.60	1.29	11.30	3.10	0.34	2.21	92.29
NPK @ 120:50:75 (T_2)	16.88	7.50	4.02	1.42	12.15	3.26	0.37	2.81	117.08
FYM @ 20 t/ha (T_3)	19.60	7.75	4.07	1.41	13.20	3.13	0.39	2.91	121.26
VC @ 6t/ha (T_4)	22.22	9.10	4.41	1.52	14.40	3.00	0.42	3.85	160.45
FYM @ 10 t/ha + 1/2 T_2 (T_5)	20.50	7.90	4.42	1.56	13.40	3.10	0.41	3.21	134.04
VC @ 3t/ha + 1/2 T_2 (T_6)	28.50	9.50	4.50	1.53	15.70	3.10	0.47	4.43	184.58
FYM @ 13.33 t/ha + 1/3 T_2 (T_7)	24.20	8.30	4.13	1.58	14.30	3.01	0.41	3.43	142.39
VC @ 4t/ha + 1/3 T_2 (T_8)	30.50	10.10	4.63	1.60	16.12	3.10	0.48	4.47	197.50
$CD_{0.05}$	**4.864**	**0.837**	**0.501**	**0.045**	**0.473**	**NS**	**0.022**	**0.358**	**9.343**

Table 9: Effect of different Treatments of Fertilizers and Manures on Soil Fertility Status after Harvest of the Crop (*Gloriosa superba*)

Treatment	*N (kg/ha)*	*P (kg/ha)*	*K (kg/ha)*	*EC(dS/m)*	*pH*
No treatment (T_1)	191.30	11.42	154.00	0.17	7.94
NPK @ 120:50:75 (T_2)	255.30	12.87	157.00	0.15	8.16
FYM @ 20 t/ha (T_3)	245.39	14.87	159.00	0.16	7.96
VC @ 6t/ha (T_4)	231.09	13.88	160.40	0.18	8.03
FYM @ 10 t/ha + 1/2 T_2 (T_5)	247.72	13.89	163.50	0.17	8.03
VC @ 3t/ha + 1/2 T_2 (T_6)	249.50	14.06	166.50	0.19	8.01
FYM @ 13.33 t/ha + 1/3 T_2 (T_7)	253.60	13.70	163.70	0.20	8.05
VC @ 4t/ha + 1/3 T_2 (T_8)	257.40	17.56	168.10	0.17	8.03
$CD_{0.05}$	5.321	1.159	1.611	NS	NS
Initial status	**251.30**	**16.10**	**162.70**	**0.15**	**7.90**

Table 10: Effect of different Tuber Size on Growth and Yield of *Gloriosa superba*

Treatments/ Weight Class	*Height (cm)*	*No. of Leaves*	*No. of Branches*	*No. of Flowers/ Plant*	*No. of Fruits/ Plant*	*Fruit Length (cm)*	*Fruit Dia. (cm)*	*No. of Seed/Fruit*	*Seed Yield/ Plant (gm)*
0-5gm (T_1)	15.00	22.14	1.00	0.00	0.00	0.00	0.00	0.00	0.00
5-10gm (T_2)	58.20	53.02	1.00	0.00	0.00	0.00	0.00	0.00	0.00
10-20gm (T_3)	62.45	57.73	1.00	3.25	1.25	2.15	1.23	10.58	0.54
20-30gm (T_4)	74.21	57.11	1.33	4.20	3.50	3.26	1.15	14.20	1.73
30-40gm (T_5)	138.64	112.8	2.00	5.75	3.85	4.21	1.24	14.13	2.51
40-50gm (T_6)	141.10	121.64	1.33	7.55	6.15	4.31	1.66	19.27	3.70
50-60gm (T_7)	140.57	144.05	2.00	8.05	5.75	4.27	1.66	17.32	3.05
>60gm (T_8)	160.11	140.11	2.00	10.75	8.33	4.13	1.58	18.85	4.25
$CD_{0.05}$	18.579	13.072	1.397	3.193	1.696	0.792	0.564	0.05	0.007

Table 11: Effect of INM on Growth and Yield of *Silybum marianum*

Fertilizers	*Sowing Dates*			
	S_1 (Oct. 15)	*S_2 (Oct. 30)*	*S_3 (Nov. 15)*	*Mean*
T_1 (NPK @ 120:100:75 Kg/ha	6.33	7.13	6.20	6.56
T_2 (FYM @20 t/ha)	5.63	6.53	5.27	5.81
T_3 (VC @ 6 t/ha+)	7.67	8.30	7.10	7.69
T_4 (FYM @10 t/ha+ ½ T_1)	8.33	9.13	8.30	8.59
T_5 (VC @ 3 t/ha+ ½ T_1)	10.27	11.77	10.03	10.69
T_6 (FYM @13.5 t/ha+ 1/3 T_1)	9.37	10.40	9.23s	9.67
T_7(VC @ 4 t/ha+ 1/3 T_1)	11.60	13.80	11.50	12.30
T_8 (Control)	4.43	5.43	4.53	4.88
Mean	**7.96**	**9.06**	**7.77**	

T_6 (IW_{40mm}: CPE_{50mm}=0.8)

T_6 (3.75t VC+75kg N+*Azoto*)

Aloe barbadensis

Spacing Trial of Sarpagandha at Agroforestry Nursery, Chatha.

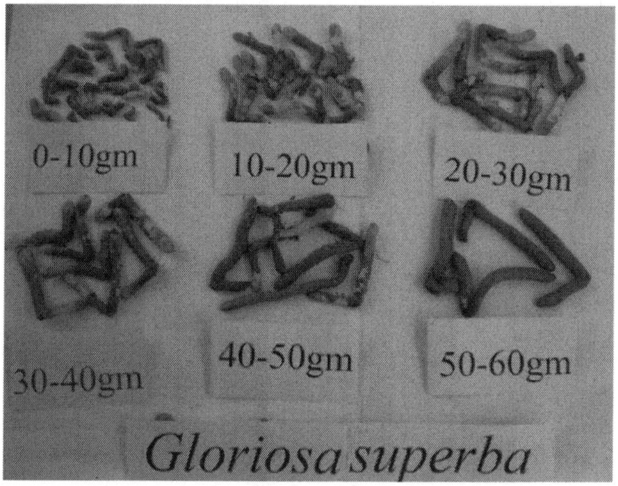

Variation in Tuber Size of *Gloriosa superba*.

Both these species (*Gloriosa* and *Rauvolfia*) figure in the prioritized list of medicinal plants enlisted by National Medicinal Plant Board, New Delhi and are found naturally in subtropical zone of Jammu division with sparse populations of their plants. To boost their cultivation it is imperative to standardize their agro-techniques.

Silybum marianum commonly known as Milk thistle or Holy thistle belongs to family Asteraceae and is native to central and southern Europe and is also found in Asia and Africa. It grows widely sub-Himalayan tract and adjoining plains of Jammu and Kashmir, Himachal Pradesh, Uttrakhand and Uttar Pradesh with an altitudinal range of 300-2400 m above mean sea level. It grows as an winter season weed on waste lands, boundaries of cultivated fields, along roads and railway tracks. Seeds of the species are used for the treatment of liver, bile related diseases, acute poisoning by mushroom, cancer, blood cholesterol and sugar.

Recommendations

Aloe barbadensis

- 50 per cent substitution of inorganic nitrogen with organic source { Vermicompost @ 3.75 t/ha or FYM @ 15 t/ha alongwith 75kg N per hectare (½ of recommended dose) and *Azotobacter*} is beneficial to increase the gel and leaf yield.
- Three accessions namely IC111271, IC112534 and IC111272 are superior over others resulting in higher gel (83.50g, 79.28g, 69.33g/plant, respectively) and leaf yield (0.96kg, 0.90kg, 0.80kg/plant, respectively).
- Irrigation of 40mm depth given at IW: CPE ratio of 0.8 is beneficial to increase the gel (0.66kg/plant) and leaf yield (1.09kg/plant).

Gloriosa superba

- For commercial cultivation of *Gloriosa superba*, tubers of >20gm weight class should be used.
- Integrated nutrient application (vermicompost @ 4 t/ha along with 40kg N, 16.67kg P and 25kg K per ha) at the time of planting (April) is advised to increase the seed yield.

Rauvolfia serpentina

- Alternate wetting and drying of seeds for two days (2 cycles) is advised to increase germination of seeds from 35.33 per cent (control) to 65.50 per cent.
- 30x 45 cm^2 spacing is optimum for cultivation of Sarpagandha which gave the highest root yield after second year of planting.

Silybum marianum

- Application of Vermicompost (VC) @4 t/ha + 1/3 of recommended dose of NPK (40:33:25 Kg/ha) and Oct. 30 sowing date, are advised for better growth and seed yield in *Silybum marianum*, under subtropical conditions of Jammu and Kashmir state

Benefit/Impact

- The agro-techniques of different medicinal species are now available for the benefit of farming community.
- Availability of planting material of different medicinal species.

AICRP-IFS

46. Diversification/Intensification: An Alternative Cropping System for Enhancing Farm Productivity and Employment under Irrigated Conditions of Jammu

Contact Scientist: **Dr. Dileep Kachroo, FSRC, Chatha**

Crop diversification is an important tool to increase the productivity and profitability vertically and play a significant role in augmenting the farm income per unit area by utilizing the fallow land for short term high value crops like vegetables. The horizontal and vertical diversification is now a days relevant under small holder production system and its having the potential for increasing the production and economics due to high cropping intensity through addition of low volume high value crops over the existing cereal-cereal based cropping system under irrigated condition of Jammu.

Methodology

An experiment has been conducted for five years at FSR Farm, SKUAST-J Chatha comprising different rice based cropping systems with varying duration of rice varieties *i.e.* Jaya (Long duration), IET-1410 (short duration) and PC-19 (Medium duration) were transplanted in Kharif season whereas during Rabi season the crops mainly Wheat, Berseem, Potato, Knol khol, Garlic, Marigold, Spinach, Broccoli, Cabbage and in summer Onion, Tomato, Cowpea, Frenchbean, Okra, Mash,

Maize+Greengram were sown in a sequence the following package of practices under irrigated conditions is recommended.

Among different rice based cropping system the five diversified system is recommended *i.e.* i) Rice-Marigold-French bean, ii) Rice-Potato-Onion, iii) Rice-Cabbage-Onion, iv) Rice-Broccoli-Mash. v) Rice-Palak-Okra are the better choice for obtaining higher net profit with B:C ratio as compared to existing Rice-Wheat cropping system the package of practices for each diversified system is given as under.

i) Rice (var. PC-19) - Marigold (Var. Pusa Narang) - French bean (var. contendor) Sequence

Agronomic Management at a Glance

	Kharif	*Rabi*	*Summer*
Management Practices	Rice (PC-19)	Merigold (Pusa Narangi)	Frenchbean (Contender)
Date of sowing	10thJuly-20th July	10thOctober to20th October	25 Feb. toIst March
Method of sowing	Transplanting	Transplanting	Line sowing
Seed rate	40 kg/ha	1 kg/ha	85kg/ha
Spacing	20cmx10cm	40cmx40cm	60cmx10cm

	Kharif	*Rabi*	*Summer*
FYM	-	8t/ha	8-10t/ha
Fertilizer ($N:P_2O_5:K_2O$)	120:60:25	120:60:25	50:100:50
Irrigation	9(no.)	4	4
Weed control	Butachlor/granules @ 5 per cent	Hand weeding	Hand weeding
Harvesting date	5-10 Oct.	15-20 Feb.	15-20 May
Cost of cultivation	17700	31745	19290
Yield, q/ha	35	42	65.5
Gross return	36750	105000	94975
Net return	19050	73255	75685

Production in terms of REY= 22.5 t/ha System net return Rs. 167990/ha

System cost of cultivation Rs. 68735/ha

System gross return Rs. 236725/ha B:C ratio = 2.90

Note: Farm gate price of Rice(PC-19) Rs. 1050/q, Marigold Rs. 2500/q, Frenchbean Rs. 1450/q

Recommendations

This sequence has been recommended as most economical which has earned net profit of Rs. 167990, system profitability of Rs. 460/ha/day and crop profitability of Rs 547/- per day, with a benefit- cost ratio of 2.90 and land use efficiency of 84 per cent. This sequence generates on farm employment for 307 days in a year.

It has also been realized that the sequence is most sustainable as no negative trends were observed in terms of soil health and the improvement of soil organic carbon content from initial value of 5.5 gm/kg soil to 6.0 gm/kg soil has been observed for a period of five years.

ii. Rice (IET-1410) – Potato (K. Badshah) - Onion (N-53) sequence

Contact Scientist:

Agronomic Management at a Glance

	Kharif	*Rabi*	*Summer*
Management Practices	Rice (IEI-1410)	Potato(K. Badshah)	Onion (N-53)
Date of sowing	10-20 July	10-20 Oct.	15-20 Jan
Method of sowing	Transplanting	Ridge sowing	Transplanting
Seed rate	40 kg/ha	2000kg/kg	10 kg/ha
Spacing	20cmx10cm	60cmx20cm	10cmx7.5cm
FYM	-	8-10t/ha	8-10 t/ha
Fertilizer (N:P_2O_5:K_2O)	60:20:20	120:60:120	100:50:50
Irrigation	9	3	3
Weed control	Butachlor granules @5 per cent	Hand weeding	Hand weeding
Harvesting date	1-5 Oct.	10-15 Jan	15-20 May
Cost of cultivation Rs/ ha	17700	46791	22415
Yield q/ha	34.88	163	140
Gross Return Rs./ha	32404	81500	121500
Net Return Rs./ha	14704	34709	99085

Production in terms of REY= 29.5 t/ha

System net return Rs. 148000/ha

System cost of cultivation Rs. 86906/ha

System gross return Rs. 235404/ha

B:C ratio = 2.12

Note: Farm gate price of Rice = 800/q, Potato = 500/q, Onion = 900/q, Cabbage= 500/q

iii. Rice (IET-1410)- Cabbage (Golden Acre)- Onion (N-53)

Contact Scientist:

Agronomic Management Practices at a Glance

	Kharif	*Rabi*	*Summer*
Management Practices	Rice (IEI-1410)	Cabbage(Golden Acre)	Onion(N-53)
Date of sowing	10-20 July	13-15 Oct	10-15 Jan
Method of sowing	Transplanting	Transplanting	Transplanting
Seed rate	40 kg/ha	0.6 kg/ha	10 kg/ha
Spacing	20cmx10cm	45cmx45cm	10cmx7.5cm
FYM	-	8-10 t/ha	8-10 t/ha
Fertilizer ($N:P_2O_5:K_2O$)	60:20:20	120:60:60	100:50:50
Irrigation	9	4	4
Weed control	Butachlor granules @5 per cent a.i.	Hand weeding	Hand weeding
Harvesting date	1-5 Oct.	28-30 Dec.	10-15 May
Cost of cultivation	17700	18701	22415
Yield q/ha	31.95	80	134.5
Gross Return Rs./ha	30060	40000	121000
Net Return Rs./ha	12360	21299	98585

Production in terms of REY= 24.17 t/ha System net return Rs. 131293/ha

System cost of cultivation Rs. 58816/ha

System gross return Rs. 191060/ha B:C ratio = 2.66

Note: Farm gate price of Rice = 800/q, Cabbage= 500/q, Onion = 900/q

Recommendation

Diversified crop sequences, rice-potato-onion and rice-cabbage-onion have also been identified as profitable and sustainable sequence for realizing higher productivity in terms of REY of 29.5 and 24.17 t/ha and net returns of Rs.1,48498 and Rs.1,31293 with B:C ratio of 2.12 and 2.66 over existing rice-wheat system with REY of 11.31 t/ha, net return of 32492 and B:C ratio (1.22), respectively under irrigated condition of Jammu. Land use efficiency under rice- potato-onion and rice-cabbage-onion was recorded 80 per cent and 84 per cent, respectively. There was no change in soil pH while soil organic carbon and available K showed slight increasing trends compared to initial values. Whereas existing rice-wheat system showed negative trend in available N, K and soil organic carbon, respectively.

iv. Rice-Broccoli-Mash

Contact Scientist:

Agronomic Management Practices at a Glance

Operation	*Rice*	*Broccoli*	*Mash*
Variety	IET-1410	Evergreen	NUL-7
Date of Sowing/ transplanting	1st week of July	3rd week of Oct	15th -30th of March
Seed rate	40 kg/ha (1-2 plant/hill)	0.6kg/ha	20 kg/ha
Spacing	15 cm x 10 cm	45 cm x 45 cm	40 cm x 10 cm
FYM	5t/ha	10 t/ha	-
Fertilizer	50:30:20	120:60:60	15:40:0
Irrigation	9-10	3	2-3
Weed Control	Butachlor (5 per cent a.i. granules) @ 30 kg/ha at 3 DAT	Hand weeding	Pendimethaline @ 400 a.i/ha
Date of harvesting	1st week of Oct	25-30th Jan	1st week of June
Cost of cultivation	Rs. 33000	Rs. 34000	Rs. 26500

Operation	*Rice*	*Broccoli*	*Mash*
Average Yield	36.1 q/ha	60.0 q/ha	9 q/ha
Gross Return Rs./ha	43320	180000	49500
Net Return Rs./ha	10320	146000	23000

Average REY (22.73 t/ha)

Average System cost of cultivation (Rs. 93500 t/ha)

Average gross return (Rs. 272820)

Average Net Return (Rs. 179320/ha)

Average System profitability (Rs. 492/ha)

Average Production efficiency (68.88)

Average Land use efficiency (90.41)

B: C Ratio (1.90)

Note: Farm gate price of Rice = 1200/q, Broccoli =3000/q, Mash =5500/q

Recommendation

This system has also been recommended as most economical and earned net profit of Rs.179320/ha, system profitability of Rs. 492/ha/day with B: C Ratio of 1.90 and land use efficiency of 90.41 with highest production efficiency of 68.88 Kg/ha/day. This sequence also generates farm employment for 330 days in a year. It has also been realized that this system is most sustainable as no negative trends were observed in terms of soil fertility besides improved soil organic carbon content from initial level of 5.5 gm/kg soil to 6.2 gm/kg soil which showed 13 per cent improvement after 5 years of study.

v. Rice-Palak-Okra

Agronomic Management Practices at a Glance

Operation	*Rice*	*Palak*	*Okra*
Variety	PC-19	All Evergreen	Lady Luck
Date of Sowing/ transplanting	1st week of July	3rd week of Oct	2nd week of March
Seed rate	40 kg/ha (1-2 plant/hill)	40 kg/ha	20 kg/ha
Spacing	15 cm x 10 cm	20 cm x 5 cm	45 cm x 30 cm
FYM	5t/ha	10 t/ha	5t/ha
Fertilizer	50:30:20	80:25:25	60:30:30
Irrigation	9-10	2-3	3-4
Weed Control	Butachlor (5 per cent a.i. granules) @ 30 kg/ha at 3 DAT	Hand weeding	Hand weeding
Date of harvesting	1st week of Oct	Last week of Feb	25-30th June
Cost of cultivation	Rs. 33000	Rs. 32500	Rs. 35000
Average Yield	40.5 q/ha	135 q/ha	105.2 q/ha
Gross Return Rs./ha	48600	121500	105200
Net Return Rs./ha	15600	89000	70200

Average REY (22.94 t/ha)

Average System cost of cultivation (Rs. 100500/ha)

Average gross return (Rs 275300/ha)

Average Net Return (Rs. 174800/ha)

Average System profitability (Rs. 478.9/ha)

Average Production efficiency (65.5)

Average Land use efficiency (95.8)

B: C Ratio (1.73)

Note: Farm gate price of Rice = 1200/q, Palak = 900/q, Bhindi = 1000/q

Recommendations

Rice-Palak-Okra has also been identified as another alternatives profitable system after Rice-Broccoli-Mash for realizing higher productivity in terms of REY of 22.94 t/ha, net return of Rs. 174800/ha and B:C Ratio of 1.73 over existing Rice-Wheat system with REY of 10.7 t/ha, net return of Rs. 74765/ha and B:C Ratio of 1.05, respectively. However land use efficiency and production efficiency under Rice-Palak-Okra was recorded 95.8 per cent and 65.5 kg/ha/day. This cropping system has also generated farm employment 350 days in a year over existing Rice-Wheat system (236 man days) in a year.

Benefits/Impact

All the above rice based diversified cropping systems gives net benefit ringing from Rs 132244/- to Rs 179320/- per hectare as compared to existing rice-wheat cropping system there by realizing at B:C ratio ranging from 1.73 to 2.90. The above recommended sequences not only increase the land use efficiency and employment generation within the farm but also are eco-friendly and sustainable as compared to rice-wheat cropping system.

47. Integrated Farming System Model towards Round the Year Production, Income and Employment Generation for Small and Marginal Farmers

Contact Scientist: Dr. Dileep Kachroo, FSRC, Chatha

Integrated farming system is a powerful tool: it holds the key for ensuring income, employment, livelihood and nutritional security in a sustainable mode for small and marginal farmers. It is a highly location specific, multidisciplinary holistic approach which represents an appropriate combination of farm enterprises *viz.* cropping systems, livestock, fishery cum poultry, forestry, horticulture intercropping with vegetable and floriculture and the means available to the farmers in such a way that the by-product of one component becomes the input of the other through the process of recycling within the system. Preservation of biodiversity, diversification of cropping or farming system and maximizing recycling ensure the success of this farming system approach toward livelihood and nutritional security for small and marginal farmers.

Methodology

A lot of efforts have been made at University level and developed an IFS model of one hectare designed for five farm family members in irrigated environment under sub-tropical zone of Jammu with integration of different profitable farm enterprises which are economically and socially acceptable to the farmer's.

Components with Land Allocation for Establishment of IFS Model of 1 ha

Different components *viz.* Crop+ horticulture (intercropped with high value vegetable crop+ floriculture)+ animal including vermin-compost and biogas, Fish cum Poultry, Mushroom, Apiary have been integrated with boundary plantation of Albizia, Grewia, Lucenia, Rajhard, Anola, Bael, popular and Karonda as bio fencing and the second row has been planted with mango trees (Amarpalli). The land allocation of one hectare IFS Model under different components of farming are bifurcated among crops including fodder block, animal, horticulture, fish cum poultry and mushroom Figure 1.

Land Allocation Designed for 5 Farmer's Family Members on the Basis of Nutritional Requirement of a Farm Family

Enterprises	*Area (ha)*
Crop including fodder	0.45
Animal	0.08

Enterprises	Area (ha)
Horticulture (Fruit + Vegetable + Floriculture)	0.37
Fish and Poultry	0.10
Mushroom	0.02
Total	**1.02**

Detail of Components of IFS Model: One ha IFS model comprising

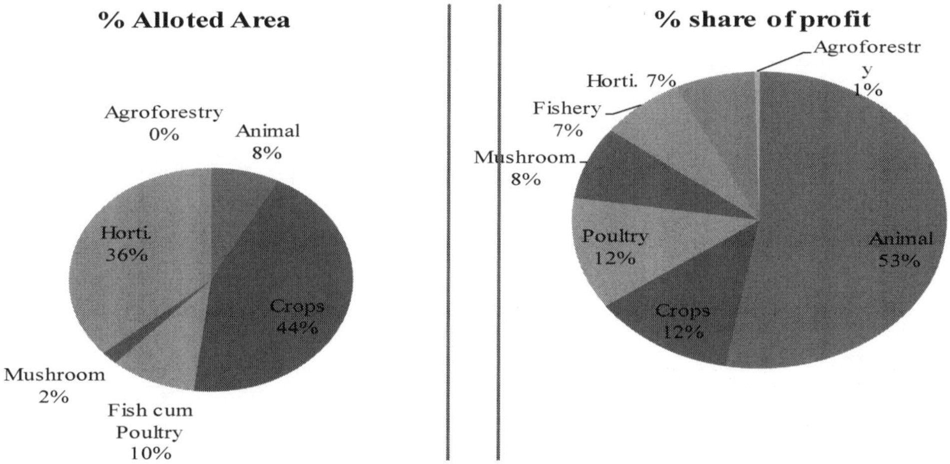

Figure 1: Area Allocated and Per cent Share of Total Profit under different Enterprises.

Cropping System (0.45 ha)

- Rice-Wheat-Green Gram
- Rice-Potato-Black Gram
- Rice-Mustard-Green Gram and
- Berseem+Oat-Maize+Sorghum with Hybrid Napier on bunds in 0.45 ha

Horticulture (0.37 ha)

- Guava (as main crop),
- Lemon (boundary crop),
- Broccoli, Knol Khol, Cabbage, Cauliflower, Okra and Marigold (as inter-crop)

Animal Unit (Dairy/Livestock 0.08 ha)

- (One Cow and One Buffalo+heifer+calves) including Bio-gas (2 m^3) and Vermi-compost (270 q capacity)

Fish and Poultry in 0.1 ha.

- Composite fish farming (Rohu, Catla, Mrigal, Common Carp and Grass Carp.) ratio 4:1.5:1.5:2:1

Mushroom (0.02 ha)

- Button (Oct. to March)- Dhingri (March to May)

Boundary Plantation

- Row-1(Outer): Albizia, Grewia, Lucenia, Rajhard, Anola, Beal, Popular and Karonda used as fruit and bio fencing.
- Row-2 (Inner): Mango trees Amarpali.

Table 1: Component-wise Average Expenditure, Gross Income and Net Profit

Component	*Net Area (ha)*	*Gross Return (Rs)*	*Total Cost (Rs/ha)*	*Net Return (Rs)*	*Per cent Share of Profit*
Crop unit					
Cropping system	0.45	75092	40201	34891	11.5
Horticulture unit					
Fruit crops	0.3	17709	6195	11514	3.8
Vegetables (intercrop)	-	31163	15663	15500	5.1
Floriculture	0.07	5365	1445	3920	1.3
Agro-forestry	-	1507	299	1209	0.4
Livestock unit					
Dairy animals (including vermicompost)	0.08	342770	192895	149875	49.2
Bio gas		15870	7380	8490	2.8
Fish cum poultry unit					
Poultry/Ducks	-	60201	25857	34343	11.3
Fisheries	0.1	29829	9129	20700	6.8
Apiary unit (03 boxes)	-	3750	2150	1600	0.53
Mushroom unit (6 qts wheat straw)	0.02	33583	10904	22681	7.4
Grand total	1.02	616839	312131	304713	100

The profit is quit comparable to existing farming system, which works out to be 0.80-0.90 lakh/ha.

Table 2: Month-wise Average Cost of Production, Gross Return, Net Return and Employment Generation under IFS Model of 1.0 ha

Month	*Total Cost Involved (Rs)*	*Gross Return (Rs)*	*Net Return (Rs)*	*Employment Generated (Man Days)*	*Production (REY kg/ha)*
July	27132	48892	21760	46.25	1877
August	26658	48634	21975	47.75	1944
September	24103	27376	3273	41.75	1092
October	24994	32467	7473	60.75	1304
November	20287	23077	2790	49.5	921
December	25857	47787	21930	48	1844
January	34653	67778	33125	54	2507
February	28201	56295	28095	48	2186
March	24689	74300	49612	50.75	2808
April	19801	33518	13717	42	1315
May	23408	60254	36846	62.25	2297
June	28656	88526	59870	44	3262
Total	308439	608904	300466	595	23357

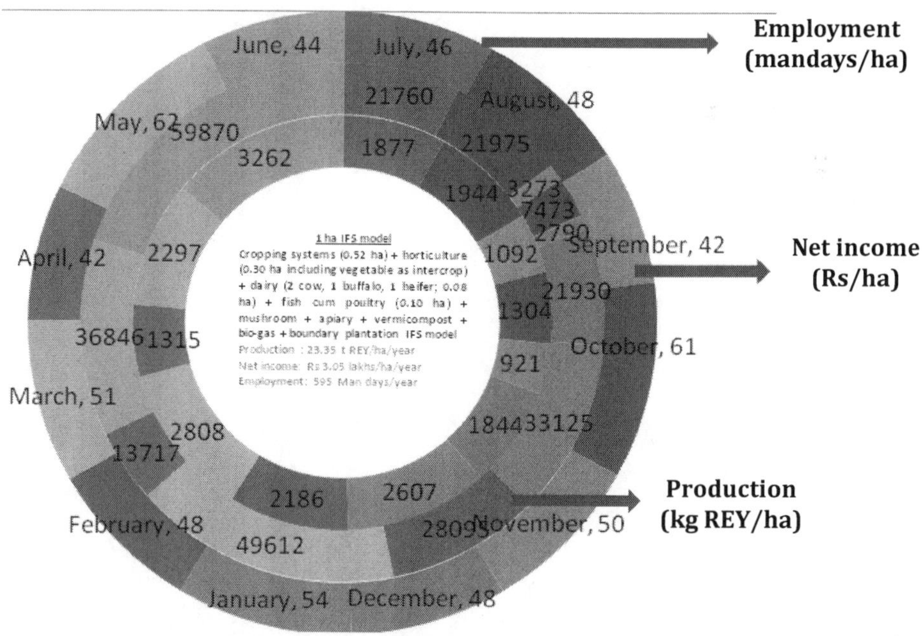

Figure 2: Integrated Farming System Model for Round the Year Production (kg REY/ha), Profit (Rs/ha) and Employment (man days/ha).

Enterprises	% profit
Crops	12.53
Animal	46.9
Hort. Fruit Veg. Flori.	 3.00 3.84 1.6
Poultry	14.6
Fish	10.02
Mushroom	7.00
Apiary	0.39
G. fodder	4.17
Ag. forestry	0.23

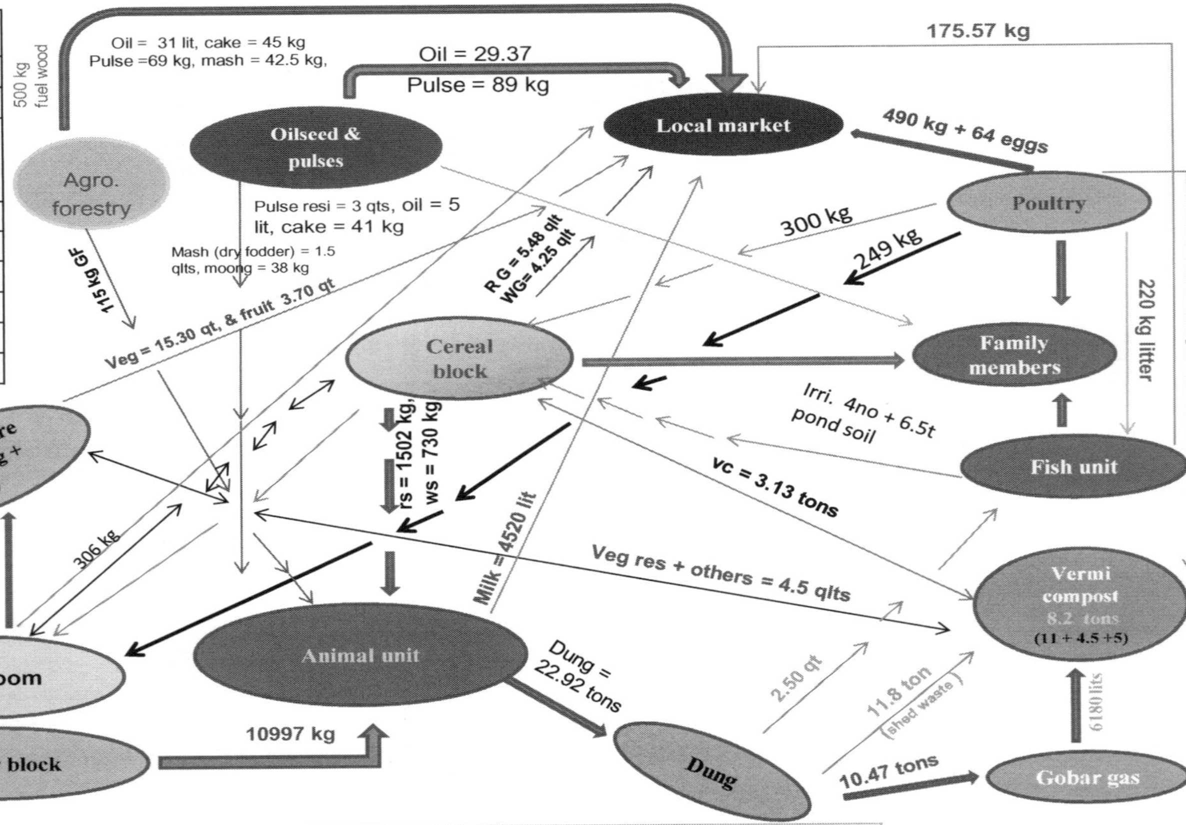

Employment generation – 595 man days/year	Profit = Rs. 3.00 lakh/ha/year

Figure 3: Recycling of Inputs and Output in the IFS Model Developed at Jammu (Jammu and Kashmir).

Table 3: Distribution of Food Commodity Production for Meeting Household Nutritional Security of Five Farm Family through Adoption of IFS Model verses Existing Farming System

Commodity	*Household Nutrition for Farm Family Member*		
	Requirement as per ICMR Standards	*Production in Existing Farming System*	*Production in Improved Farming System*
Cereals (kg)	1320	3895	1370
Pulses (kg)	150	0	78
Oilseeds (kg)	150	20	80
Potato (kg)		0	800
Vegetables (kg)	768	50	1157
Fruits (kg)	270	50	823
Milk (litres)	660	1200	5648
Meat and Fish (kg)	66	0	476
Egg (nos.)	900	0	92
Green fodder (t)	26.8	3.5	14.3
Dry fodder (t)	6.25	6.02	2.77
Mushroom (kg)		0	361
Turmeric (kg)		0	44
Agro forestry (kg)		0	500
Rice equivalent Yield (t)	6.3	23.35	

Performance of Basmati-370 under IFS Model

Wheat Crop in Rice-Wheat Cropping System under IFS Model

Performance of Mustard Crop after Rice under IFS Model

Horticulture Component under IFS

View of IFS Model

Mango Plantation under IFS Model

Recommendation

One ha Integrated Farming System Model comprising of cropping systems (rice-wheat-green manuring, rice-potato-blackgram, rice-mustard-greengram and berseem+oat-maize+sorghum with hybrid napier on bund) in 0.45 ha + horticulture (Guava as main crop, Lemon and mango (Amarpali) as boundry crop and brocolii, KnolKhol, Cabbage, Cauliflower, okra and Marigold as intercrops) in 0.37 ha +

Backyard Poultry

Hon'ble Dy. CM, J&K Interacting with Hon'ble VC, SKUAST-J during his Visit IFS Model

Two Cows and One Heifer under Animal Component of IFS

Four Tank Vermicompost unit under IFS Model

dairy (2 cow, 1 buffalo heifer) including 2 m^3 bio-gas and vermicompost unit in 0.08 ha + fish cum poultry in 0.1 ha) + mushroom (button and dhingri) in 0.02 ha developed for the Mid to high altitude plain zone (JK-1) in Western Himalayas provides *round the year* average production (23.35 t REY/year), profit (Rs 3.05

Biogas Unit under IFS Model

Netting of Fish under IFS Model

lakhs/year) and employment (595 man days/year). Among all the enterprises the highest average net profit of Rs. 149875 (contributing 49.2 per cent share of profit) was realized from Animal Unit having two Cows and One Buffalo heifer including Vermicompost/farm waste recycling component then followed by crop block including green fodder (Rs. 348881), Poultry (Rs. 34343), Mushroom (Rs. 22681) and Fisheries (Rs. 20700) which contributes profit share of 11.5, 11.3, 7.4 and 6.8 per cent respectively. The overall average net profit Rs. 304713/year/ha was quite comparable to existing farming system which works out to be 0.80 to 0.90 lakh/ha (Table 1).

The maximum average production and profit was realized in June (Table 3, Figure 1) while employment was in May month signifying the work even during lean period. The model also meets around 75 per cent of inputs required for different enterprises within the farm besides providing all the commodities (cereals, pulses, oilseeds, vegetables, fruits, mushroom, milk, egg, and fish) required for the farm family (Figure 2). The IFS Model of 1 ha contributed 60 kg Nitrogen, 20 kg P_2O_5 and 40 kg K_2O per year through recycling/vermicomposting besides bio gas unit of $2m^3$ generated bio gas (@ 0.8 kg LPG/day/15 kg dung) equivalent to 17 LPG Cylinder of 14.5 kg capacity, respectively. Moreover this IFS model under sub-tropical condition of Jammu region is having the potential to improve the farm income to three folds besides will address sustainability issues in the long run.

Not withstanding with the doubling the farm income and generating employment through IFS it has been realized that model developed have the capacity to meet the household nutritional requirements of five farm family members. From the perusal of the data Table 3, it was observed that a farm family on an average using IFS model generates different commodities which comparatively balanced and higher than the existing farming system.

AICRP-Rice

48. System of Rice Intensification – Efficient Input Use Methodology

***Contact Scientist:* Dr. Anuradha Saha**

"Rice grown under irrigated condition is facing the threat of water shortage. This is forcing a paradigm shift towards maximizing output per unit of water instead of per unit of land."

The System of Rice Intensification basically refers to a particular set of practices which improve plant health and yield. It involves a set of practices for plant, soil, water and nutrient management. The cultivation of rice under SRI involves certain principles aimed at lesser seed, 8-12 days transplanting of rice seedlings, more use of organic manures/integration of organics and inorganics, promotes soil biotic activities in and around the root zone, weeding through conoweeders and no use of insecticides as well as improving aeration in fields through alternate wetting and drying cycle, thereby saves irrigation water.

Detail Methodology of SRI

Raising Seedlings and Nursery Management

- Area: 100 m^2 nursery area is required for one hectare area.
- Prepare nursery bed of 4 X1.25 m^2 size with FYM and soil mixture of 2:1 ratio for one kanal area

- Nursery bed should be raised 12.5-15 cm in height along with bunds and channels. The bed should be prepare near the main field
- 400 g per kanal rice seed which should be pre-germinated, should be sown through broadcasting and cover with FYM and soil mixture
- Irrigate the beds twice daily with rose cane.

Preparation of Main Field, Nursery Uprooting and Transplanting

- Field should be leveled, puddle and standing water should be drained out before transplanting.

Nursery Uprooting with Soil

Transplant Single Seedling at 25 X 25 cm

- Uniform spacing is required for easy weeding and hoeing. Use any rope marked with 25 cm distance for facilitating square transplanting of seedlings @ 16 seedlings per square m.
- Transplant 8-12 days old seedlings (2-3 leaf stage). Nursery should be uprooted through steel plate by inserting below the seed bed to get seedlings intact with seed and soil without damaging roots.
- Single seedlings with soil should be transplanted gently at shallow depth at 25 cm distance. Interval between uprooting and transplanting should be minimum (not beyond an hour). This kind of transplantation resulted in quick recovery and establishment of plant.

Water Management

- Subsequent irrigations (2 -2.5 cm) were given when soil developed fine hair line cracks
- Maintain shallow submergence (2.5 -3 cm) from panicle initiation up to 15 days before harvesting
- However, there should be a thin film of standing water while operating the weeder and same water is retained for weed incorporation

Nutrient Management

- Apply half of the recommended N and total P and K as basal before puddling and incorporate. Apply the second dose (25 per cent) of N at the time of 2nd weeding (20 DAT) and final dose of 25 per cent N a week before panicle initiation.
- SRI method encourages application of organic manures as it enhances the macro and micro nutrient content in the soil. Different sources of organic manures should be use rather than single source. It also helps in suppressing population of harmful plant parasitic and promoting buildup of beneficial microbial.
- Quantity of organic manures can be adjusted in such a way to supply recommended dose of N through different sources of organic manure.

Weed Management

Cono Weeders

- Weed growth is more in SRI due to wider spacing and alternate wetting and drying. Conoweeding helped in replenishing the nutrients through incorporation of weeds, increased aeration and microbial activity thereby improving plant as well as soil health.
- First weeding with shallow standing water at 10-12 days after transplanting and a minimum 3 weedings at 10-12 days interval is required

Cono Weeding in Standing Water

After Cono Weeding

Status of SRI in Jammu

SRI was initiated in *kharif*, 2005 at Research Farm of SKUAST-J, Chatha through AICRIP (Rice) research trial from Directorate of Rice Research, Hyderabad. The positive results emanated from SRI research trial acted as fillip and two externally funded projects were completed at six different locations/villages in Jammu district, with funding from WWF-ICRISAT, Hyderabad in *kharif*, 2007 and *kharif*, 2008. The broad outline and the achievements of the externally funded projects on SRI are given below.

Title of the Project	*Achievements*
Demonstration and on the spot training of System of Rice Intensification (SRI) on farmers fields' of Jammu.	☆ SRI was demonstrated with all the principles. ☆ Five farmers were selected and imparted on the spot training at respective demonstration sites of farmers' choice. ☆ Seed rate of 8 kg/ha was standardized. ☆ Seedling age of 12-14 days was standardized. ☆ Insitu green manuring/FYM/other organic sources (50 per cent of recommended N) + 50 per cent recommended fertilizer dose was standardized. ☆ Results showed 49 per cent average increase in rice yield (Cultivar PC-19, a local HYV) under SRI as compared to farmer's practice at respective sites. ☆ One field day on SRI was organized and attended by Hon'ble Vice Chancellor, SKUAST-Jammu, staff from Agriculture Dept. J&K, field functionaries and about 150 farmers.
Scaling up the productivity of rice through SRI as an alternative option for sustainable natural resource management in Jammu.	☆ In second year, another project was implemented at farmers field on cluster basis demonstration in village Mota, R.S. Pura, Jammu. ☆ Five trained farmers acted as master trainers. ☆ Two master trainers attended 2nd National Symposium on SRI at Agartala, Tripura. ☆ SRI was upscaled to 26 acres area.

Title of the Project	*Achievements*
	☆ SRI was demonstrated with practices standardized such as ❖ Seed rate @ 8 kg/ha ❖ Insitu green manuring+50 per cent RFD (30:20:10) ❖ Transplanting of 12 -14 days age seedling ❖ Conoweeding twice in single direction was conducted ❖ Major demonstrated area was under Basmati-370. Results obtained recorded 22.5 per cent average increase yield in Basmati-370. ❖ One day workshop on SRI was organized and attended by Agricultural Production Commissioner, Govt. of J&K, Hon'ble Vice Chancellor, SKUAST-Jammu, Directors from Agriculture Dept, Jammu. J&K, field functionaries and about 250 farmers.

Results at Farmers' Fields with SRI

Name of the Village	*Yield Increased over Conventional Method (per cent)*
Chatha	48 per cent
Khandwal	45 per cent
Giddar Gallian	50 per cent
Badyal Brahmana	50 per cent
Mota	60 per cent
Data from Demonstrations (Kharif 2007) Cultivars: PC-19 and Sharbati	

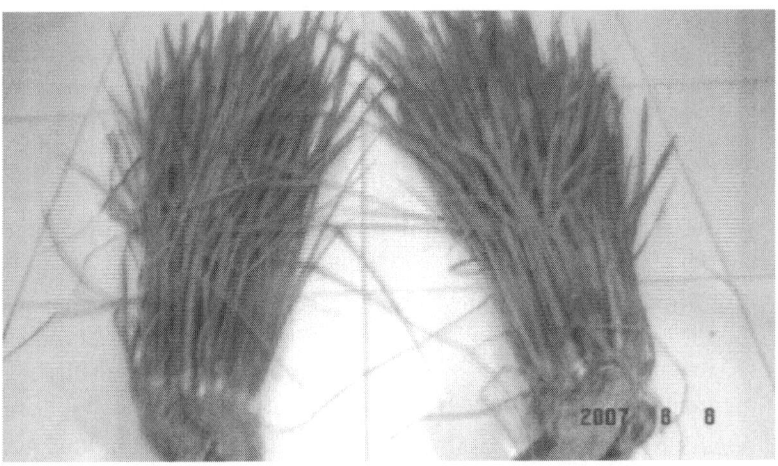

Profuse Tillering (PC-19)

Water Saving in SRI

Name of the Village	*Water Used in Conventional Method*	*Water Used in SRI*	*Water Saved (per cent)*
Chatha	14877 ($m^3 ha^{-1}$)	10565 ($m^3 ha^{-1}$)	28.98
Khandwal	15235 ($m^3 ha^{-1}$)	11345 ($m^3 ha^{-1}$)	25.53
Giddar Gallian	15753 ($m^3 ha^{-1}$)	11423 ($m^3 ha^{-1}$)	27.49
Badyal Brahmana	15802 ($m^3 ha^{-1}$)	11654 ($m^3 ha^{-1}$)	26.25
Mota	15487 ($m^3 ha^{-1}$)	11395 ($m^3 ha^{-1}$)	26.42
Irrigation + rainfall = total water used			

Recommendation

Sl.No.	*Items*	*Conventional Method*	*SRI Method*	*Per cent of Saving*
1	Seed	30 kg (Basmati)40 kg (Course)	8 kg	60-80 per cent
2	Weedicide	30 kg butachlor	No weedicide	100 per cent
3	Irrigation water	14000 -16000 m3/ ha	9000- 11500	25-30 per cent
4	Time	All varieties mature as per their lifecycle	All varieties are mature 7-10 days early than conventional method	7-10 days
5	Yield	Yield recorded as per their potential of all varieties	All varieties showed positive result	25-45 per cent (Basmati and non basmati)

- SRI cannot be adopted everywhere. It may not be suitable in command area where water release is highly uncertain and in low land under uncontrolled water situations especially during *kharif* season.
- SRI being the skilled and knowledge intensive methodology requires complete involvement of the farmers. As the marginal farmers may involve all the action of SRI the suitable beneficiary of SRI should be marginal farmers.
- In SRI method encourage application of organic manures as it enhance the macro and micro nutrient content in the soil, protects the soil from degradation and other hazardous effects. The availability of organic manure in large quantity may help one of the principles of SRI.
- Weeding is one of the major constraints in adoption of SRI. Availability of cono weeder is a compulsory component, as it helps not only for

weeding it incorporate the weeds in the soil to increase organic matter and aeration.

- SRI is most suitable for seed multiplication as initial seed requirement is small and rouging is easy to planting of single seedling and wider spacing.

Workshop on SRI Conducted for Farmers, Agril Dept and Scientist from SKUAST-Jammu.

SRI Field Visit by farmers, Agril Dept and Scientists from SKUAST-Jammu.

SRI Field Visit by farmers, Agril Dept and Scientists from SKUAST-Jammu.

AICRP-Weed Management

49. Weedy Rice Management in Transplanted Rice

Contact Scientist: Dr. B.R. Bajaya

In Jammu region area under Rice is about 114.00 thousand ha and the production was about 328.4 thousand tones. The rice production in Jammu region is major commercial activity, especially in irrigated subtropics. In recent years, the rice production especially in subtropical plains of Jammu division is encountering a new biotic threat- of the Weedy Rice. The weedy rice may be defined as a weedy population of genus *Oryza* that (1) is similar to cultivated rice but has greater seed dormancy, susceptibility to seed shattering; (2) grows vigorously in and around cultivated rice; (3) is very diverse phenotypically and genotypically; (4) is highly competitive and difficult to control; and (5) reduces farmers income by reducing grain yield, quality and profit.

Table 1: Effect of different Treatments on Weedy Rice Density, Weedy Rice Dry Matter, Yield and Economics (Pooled data of 2 years)

Treatments	*Weedy Rice Density (m^{-2})*	*Weedy Rice Dry Matter ($g\ m^{-2}$)*	*WEC*	*Grain Yield (q/ha)*	*Net Return (₹/ha)*	*B:C Ratio*
Stale seed bed with Glyphosate @1.5 kg/ha	2.25 (4.16)	4.66 (21.95)	78.39	29.50	64286	1.93

Treatments	*Weedy Rice Density (m^{-2})*	*Weedy Rice Dry Matter ($g\ m^{-2}$)*	*WEC*	*Grain Yield (q/ ha)*	*Net Return (₹/ha)*	*B:C Ratio*
Stale seed bed with Paraquat @ 0.8 kg/ha	2.38 (4.79)	5.06 (26.66)	73.75	28.73	61582	1.83
Recommended Practice	3.70 (12.73)	9.07 (81.89)	19.39	20.96	40795	1.40

Data was subjected to square root transformation gokiddd X+1. Original values are in parenthesis

Weedy Rice Infested Fields

Germination of Weedy Rice in Tilled and Untilled Field

Treatment Application

Effect of Treatment

Rice Field Free from Weedy Rice

Methodology

1. Irrigate the field after harvest of Rabi sown crop
2. Plough the land when it comes to optimum moisture and allow the surface to germinate
3. Apply glyphosate 1.5 kg/ha or paraquat 0.8 kg/ha with 500 L of water on germinated weedy rice plants.
4. This process must be completed well before 10-15 days of transplanting.

Recommendation

The stale seed-bed with glyphosate 1.5 kg/ha or paraquat 0.8 kg/ha have been found as most effective and economical method for control of the weedy rice in transplanted rice

Benefit/Impacts

Basmati rice is grown on more than 60000 hectors under irrigated sub-tropics of Jammu region, most of which is prone to menace of weedy rice. If the proper weedy rice management strategy is followed *viz.* stale seed-bed with glyphosate @ 1.5 kg/ha or paraquat @ 0.8 kg/ha before transplanting, the total expenditure on management of this weed would be Rs. 4500/ha and in response to this an enhancement of about 4-5 q/ha crop yield can realized which may fetch an additional returns of about Rs. 20000/ha over existing recommended practice.

AICRP-Rapeseed and Mustard

50. Moisture Stress Mitigation in Indian Mustard using Agrochemicals

Contact Scientist: **Dr. Rajeev Bharat**

The abrupt increase in average temperature during the vegetative as well as reproductive stages of crop sometimes leads to forced maturity thus shortening the life cycle of the crop. This results in significant reduction of Indian mustard yield to an extent of 60-70 per cent. To overcome this yield loss small interventions such as spray of agrochemicals *viz.* thiourea, potassium nitrate and urea in low concentrations can help in overcoming moisture deficit as well as deficiency of essential nutrients at critical stages of crop growth.

Methodology

Moisture stress for an extended period (upto 50 DAS) forces the crop to initiate flowering early by reducing the vegetative phase. To prolong the vegetative phase use thio-urea @ 0.05 per cent at 50 per cent flowering stage by mixing 0.5 g in one litre of water (7.5 grams in 15 litres pump) using knapsack sprayer with flat fan T-jet nozzle using a spray volume of 700 litres ha^{-1} of water followed by another spray at 50 per cent siliquae filling stages in case of prolonged dry spell which coincides between 50-60 and 90-100 days after sowing respectively. Foliar application of thio-urea @ 0.05 per cent at 50 per cent flowering stage followed by 50 per cent siliqua filling stage can increase the crop yield to an extent of 15-38 per cent depending on the prevailing conditions thus reducing the crop yield losses significantly.

Table 1: The Performance of Thiourea @ 0.05 per cent on Yield of Indian Mustard

Sl.No.	*Treatment*	*Yield ($q\ ha^{-1}$)*	*Per cent Increase (Over farmers practice)*
1	No application (Farmers practice)	6.22	-
2	1 per cent urea spray at 50 per cent flowering stage	7.77	24 per cent
3	0.05 per cent thiourea spray at 50 per cent flowering stage followed by 50 per cent siliqua formation stage	8.61	38 per cent

View of Crop with Farmers Practice

View of Crop Sprayed with 0.05 per cent Thio-urea

Recommendation

Foliar application of 0.05 per cent thio-urea spray at 50 per cent flowering stage followed by 50 per cent siliquae filling stage was found to be the most economical treatment in mitigation of moisture stress and resulted in significant increase in seed yield of Indian mustard (*Brassica juncea*) crop than other treatments in comparison under rainfed conditions of Jammu region.

Benefits/Impact

During the last few years prolonged dry spell and heat stress are recurring phenomenon leading to moisture stress at flowering and siliqua formation stages of mustard. This abiotic stress causes approximately 50-70 per cent yield loss. This amounts to about 4-5 quintal per hectare reduction in crop yield, thus causing a huge loss to the farmers income. This adds to the import bill of oilseeds on the government exchequer which is estimated to around Rs 17-20 crores annually in Jammu region. Hence, adoption of this technology will help in increasing the farmer's income and reducing the dependence on imports from other states and country significantly under prolonged dry spell and moisture stress condition.

AICRP-PRSS, Samba

51. Bio-efficacy of different Herbicides for Broad Spectrum Weed Management in Chickpea

Contact Scientist: **Dr Brij Nandan**

Methodologies

The experiment was conducted for the two consecutive Rabi seasons of 2013-14 and 2014-15 under RBD design.

Result

Data with significant result among the nine herbicidal treatments, the application of Pendimethalin 30 EC formulation + Imazethapyr 2 per cent (Ready mix combination)* @ 1.0 kg/ha PE + one hoeing at 30-35 DAS registered significantly higher grain yield of chickpea (1473.00 kg/ha) over all other treatments followed by Pendimethalin 30 EC formulation + Imazethapyr 2 per cent (Ready mix combination) @ 1.0 kg/ha PE with 1269 kg/ha grain yield of chickpea. The lowest grain yield of chickpea (499 Kg/ha) was recorded with treatment one HW at 30-35 DAS. *Chenopodium album, Melilotus indica, Fumeria parviflora, cyperus rotundus, Cynodon dactylon, Phalaris minor, Anagalis arvensis, railway creeper* were the common weed flora found in the experimental site.

Note: No phototoxic effect on crop was observed during the course of investigation.

Significant observation noted during the course of investigation that the appearance of railway creeper weed is not found in the plots where Pendimethalin 30 EC formulation + Imazethyper 2 per cent (Ready mix combination @ 1.0 kg/ ha) followed by 1HW was used however, it was observed in all the plot where rest of the herbicides were used. Moreover, this is very important information as the appearance of this weed generally seen in between end of February to March and its spread is so fast in the chickpea fields that it is almost impossible to control this weed at this point of time as the crop -is approaching towards its maturity stage

Treatments	*Grain Yield (Kg/ha)*		*WCE (per cent)*	
	2013-14	*2014-15*	*Mean*	*Mean of 2 Years*
Pendimethalin EC formulation (0.75 - 1.0 kg/ha), PE + one hand weeding at 25 -30 DAS	**723**	**986**	**854.5**	**68.3**
-Pendimethalin CS formulation (0.75 - 1.0 kg/ha), PE	786	873	829.5	34.5
Pendimethalin CS formulation (0.75 - 1.0 kg/ha), PE + one hoeing at 30-35 DAS	973	969	971	53.1
Oxyflurofen @ 0.25 kg/ha PE at 20 DAS + one hoeing at 30-35 DAS	588	781	684.5	32.1
Fenoxyprop ethyl @ 60 g/ha POE at 25 - 30 DAS	789	622	705.5	28.6
Pendimethalin 30 EC formulation + Imazethapyr 2 per cent (Ready mix combination) @ 1.0 kg/ha PE	1269	958	1113.5	46.5
Pendimethalin 30 EC formulation + Imazethapyr 2 per cent (Ready mix combination)* @ 1.0 kg/ha PE + one hoeing at 30-35 DAS	1473	1331	1402	88.6
Pendimethalin CS formulation (0.75 - 1.0 kg/ha, PE + Imazethapyr 2 per cent (tank mix combination @ 1.0 kg/ha)	1088	944	1016	35.5
One hoeing/hand weeding at 30-35 DAS	499	740	619.5	31.2
CD at 5 per cent	169	330	249.5	—

Pendimethalin 30 EC formulation + Imazethapyr 2 per cent (Ready mix combination)* @ 1.0 kg/ha PE + one hoeing at 30-35 DAS registered a grain yield of 1402 Kg/ha significantly higher grain over all other treatments followed by Pendimethalin 30 EC formulation + Imazethapyr 2 per cent (Ready mix combination) @ 1.0 kg/ha PE with 1016 kg/ha grain yield of chickpea. The per cent increase of the best treatment over recommended practice was 64.07 per cent with weed control efficiency of 88.6 per cent.

Recommendations

Pendimethalin 30 EC formulation + Imazethapyr 2 per cent (Ready mix combination)* @ 1.0 kg/ha PE followed by one hoeing at 30-35 DAS in chickpea crop is beneficial in controlling the weeds.

Benefits/Impact

Weeds are the serious threat in rainfed areas of Jammu region which hinder the promotion of chickpea cultivation in rainfed areas of Jammu region. Chickpea being a main pulse crop of *kandi* areas of Jammu region therefore, this technology will certainly helpful in promoting and increasing the chickpea production as well as area under *kandi* areas.

AICRP-IWM, Chatha

52. Laser Leveling for Improving Yield and Saving Water in Rice-Wheat Sequence under Irrigated Conditions of Jammu

***Contact Scientist:* Dr. Vijay Bharti**

Methodology

Rice- wheat sequence is the important one providing food security both at national and regional scale. Precise leveling can be attained with laser land leveler which can save water, increases use efficiency of inputs and can improve farm income through improved productivity. Laser levelling aims at smoothening the land surface from its average elevation using laser equipped drag buckets to achieve precision in land leveling. With the advent of Laser Leveler, the objectives of land smoothening, water use as well as its efficiency and economic benefits can be achieved.

Table 1: Effect of Laser Leveling on Grain Yield and Water Use Efficiency in Rice-Wheat Sequence during 2011-2013

Treatment	*Rice-Wheat Sequence*					
	Grain Yield (kg/ha)	*Rainfall (mm)*	*Irrigation Water Applied (mm)*	*Total Water Applied (mm)*	*WUE (kg/ha-mm)*	*B:C Ratio*
Farmer levelled	4790	1173.2	580	1703.2	8.37	2.88
Laser levelled	5837	1173.2	460	1678.2	10.32	3.22

Table 2: Effect of Laser Leveling on Grain Yield and Water Saving in Rice-Wheat Sequence on Farmers' Fields during 2015-2017

Crop Sequence	*Yield (q/ha)*		*Yield Increase (per cent)*	*Irrigation Water Applied (m^3/ha)*		*Water Saving (per cent)*
	Farmer Levelled	*Laser Levelled*		*Farmer Levelled*	*Laser Levelled*	
Rice-wheat	70.1	80.2	14.3	198	148	25

Rice-Wheat Sequence in Laser levelled Plots.

Recommendation

The stakeholders of Jammu like Department of Command Area Development, Department of Agriculture must promote laser leveling practice on farmers' fields falling under irrigated area as it improved productivity of rice-wheat sequence by 21.8 per cent and upscaled WUE to 10.32kg/ha-mm as compared to farmer's traditionally leveled field(8.37kg/ha-mm). Benefit: cost ratio in laser leveled plots improved to 3.22 as compared to 2.88 in traditionally leveled plots. Additional returns of Rs. 32000/ha can be realized by incurring expenditure of Rs. 8500/ha on laser leveling.

Impact

The technology is moist suited for irrigated canal commands where rice-wheat is prevalent cropping sequence. The technology was successfully demonstrated to farmers of village Sultanpur and Haripur Sainia of Kathua district, Jammu and Kashmir. Laser leveling is now a common practice among the farmers and it is available on custom hiring basis in different pockets of Jammu.

53. Multiple Water Use Model for Waterlogged Areas of Jammu Region

***Contact Scientist*: Dr. Vijay Bharti**

Objectives

1. To bring waterlogged areas under cultivation
2. To identify productive and economic farming system using raised/sunken bed technology
3. To improve the livelihood of the farmers

Location

Physical model (4.0 ha) at SKUAST-J, Chatha

Initial Condition of the Area

The initial condition of the water-logged/water stagnant area measuring 4.0 ha is given in figure '1' and '2'

Final Condition of the Developed Area

The major components of the area developed as physical model is given hereunder as presented in figure '3':

1. Agro-forestry- Eucalyptus plantation on sloppy side 1.35 ha
2. Raised beds each of 80 m x 9 m size: **(06 nos.)** 1.25 ha
3. Sunken beds/Pond area/water body 80 m x 9 m x 1.5 m: **(06 nos.)** 0.85 ha
4. High density guava plantation on raised beds: 0.55 ha
   (Adjoining to raised beds at sr. no. 2 and 1)

Crop Diversification

Raised Beds

a. Cereal- based cropping system
b. Vegetable/Exotic vegetable- based cropping system
c. Floriculture- based cropping system
d. Pulses- based cropping system
e. Horticulture (High density Guava)

Sunken Beds

Fisheries (Indian major carps- Catla, Rohu, Mrigal and Exotic carps- Silver carp, Grass carp and Common carp)

Outcome

The results of the study would be applicable to about 20,000 ha of waterlogged areas within different districts of Jammu province and can be scaled up in Kashmir province as well.

Impact

- Enhanced productivity of wet lands
- Enhanced water productivity
- Enhanced income of farming community

Initial Condition of the Waterlogged Area.

Final View of the Physical Model on Multiple Use of Water for Crop Diversification within Waterlogged Areas of Jammu Province.

54. Increased Potato Production and Water Saving through Sprinkler Irrigation in Canal Command of Jammu

Contact Scientist: **Dr. Vijay Bharti**

Methodology

Potato is predominant vegetable crop after rice and wheat in canal commands. Over irrigation during rice delays field operations in other succeeding crops. Potato is more sensitive to soil water deficits than rice and wheat and requires adequate and consistent soil water during most of its growth period. A workable and efficient irrigation management strategy for potato requiring light and frequent irrigations is required and sprinkler irrigation can be one of them. Sprinkler irrigation systems are capable of applying light, frequent, and uniform irrigations during the growth of potato. A study was conducted at research farm of SKUAST-Jammu during *rabi* 2011-12 and 2012-13 to compare different irrigation schedules under different irrigation methods. Pooled data of two years recorded at par potato tuber yield with sprinkler and skip furrow irrigation methods but significantly higher over flooding method. The ncrease in tuber yield with sprinkler method was 18.75 per cent as compared to flooding method. On the basis of mean data of two years, sprinkler method saved 38.6 per cent and 20 per cent water over flooding and skip furrow methods. Also, sprinkler method recorded higher water productivity and produced 38.24 kg potato/mm water as compared to 22.73 kg potato/mm water with flooding and 32.45 kg potato/mm water with skip furrow method. In both the years of experimentation, rabi 2011-12 and 2012-13 scheduling irrigation at 0.3 PE resulted in water saving of 22 per cent over 0.7 PE and 11 per cent over 0.5 PE. Scheduling at 0.3 PE recorded higher water use efficiency and produced 34.27 kg potato/mm of water as compared to 0.7 PE (26.93 kg potato/mm water) and 0.5 PE (30.30 kg potato/mm water).

Table 1: Effect of Irrigation Methods and Scheduling on Potato (Pooled data of 2011-12 and 2012-13)

Irrigation Methods	*Yield (t/ha)*	*Irrigation Water Applied (mm)*	*Rainfall (mm)*	*Total Water Applied (mm)*	*Water Use Efficiency (kg/ha-mm)*	*B:C Ratio*
Flooding	13.60	199.9	398.2	598.1	22.73	1.79
Sprinkler	17.15	50.2	398.2	448.4	38.24	2.64
Skip Furrow	16.15	99.4	398.2	497.6	32.45	2.63

Irrigation Methods	Yield (t/ha)	Irrigation Water Applied (mm)	Rainfall (mm)	Total Water Applied (mm)	Water Use Efficiency (kg/ha-mm)	B:C Ratio
CD (5 per cent)	2.65					
Irrigation scheduling						
0.3PE	17.15	102.2	398.2	500.4	34.27	2.76
0.5PE	15.60	116.5	398.2	514.7	30.30	2.43
0.7PE	14.25	130.8	398.2	529.0	26.93	2.00
CD (5 per cent)	1.67					

Sprinkler Irrigation in Potato Crop

Recommendation

The stakeholders of Jammu like Department of Command Area Development, Department of Agriculture must promote sprinkler irrigation in potato under irrigated conditions as it increased potato yield from 136 q/ha to 171 q/ha which is 35 quintals more over farmer method of flooding. Application of water through sprinkler saved 75 per cent of irrigation water applied over flooding and increased the water use efficiency from 30.47 kg/ha-mm under flooding to 58.4 kg/ha-mm under sprinkler irrigation. Benefit cost ratio of 2.64 can be achieved by adopting sprinkler irrigation as compared to 1.79 in flooding method in potato.

Impact/Benefit

The technology demonstrated in village Rattian near Miransahib of district Jammu proved that 29 per cent more tubers can be harvested with sprinkler irrigation and 48 per cent irrigation water can be saved. This technology if promoted in 10 per cent of area under potato cultivation which is 579 ha in Jammu region can raise production by 2000quintals and can save 86420m^3 of irrigation water.

AICRP-Wheat and Barley

55. Management of Powdery Mildew Diseases of Wheat

Contact Scientist: **Dr. M.K. Pandey**

Powdery mildew is a fungal leaf disease that reduces yield and grain quality in susceptible wheat varieties. It can be very hard to control once it is established. Infection early in the season can significantly reduce yield (by up to 25 per cent) by reducing photosynthetic leaf area but more crucially by affecting yield potential through causing abortion of young tillers. Severe infection of powdery mildew at early growth stages can also cause stunting of plants. Infection later in the season, such as between stem elongation and flowering stages, can reduce photosynthetic leaf area and cause reduced grain size, lower yields and quality downgrades. Severe infection can also cause crop lodging through weakened stems.

Methodology

Treatments: Following treatments were used in the experiment:

- **T1 :** One spray of Tilt 35 EC @ 0.1 per cent at initiation of disease
- **T2 :** Two spray of Tilt 35EC @ 0.1 per cent (Ist spray at initiation of disease and 2nd spray after 15 days)
- **T3:** One spray of Bavistin 50@ 0.1 per cent at initiation of disease
- **T4:** T2 : Two spray of Bavistin 50 per cent @ 0.1 per cent (Ist spray at initiation of disease and 2nd spray after 15 days)

Table 1: Management of Powdery Mildew of Wheat and Yield

Treatment	*Rate (per cent)*	*Disease Severity (per cent)*		*Mean Severity (per cent)*	*Grain Yield (q/ha)*		*Mean Grain Yield (q/ha)*	*Yield Increased (q/ha)*
		2009-10	*2010-11*		*2009-10*	*2010-11*		
One spray of Tilt 25 EC	0.1	2.66	2.36	2.51	41.70	42.45	42.07	7.45
Two spray of Tilt 25EC after 15 days interval	0.1	0.27	0.28	0.27	44.73	44.41	44.57	10.48
One spray of Bavistin 50WP	0.1	5.94	6.04	5.99	37.03	35.60	36.31	2.78
Two spray of Bavistin 50 per cent WP after 15 days interval	0.1	5.49	5.82	5.65	38.28	37.08	37.68	4.03
Three spray of Sulfex after 10 days interval	0.3	5.49	5.32	5.40	39.60	39.39	39.49	5.35
Two spray of Karathane 350EC after 15 days interval	0.05	4.10	4.04	4.07	40.40	40.55	40.47	6.15
Control (No spray)	-	6.99	7.33	7.16	34.25	33.91	34.08	-

- **T5 :** Three spray of Sulfex @ 0.3 per cent (Ist spray at initiation of disease, 2nd spray after 10 days and 3rd spray after 10 days of 2nd spray)
- **T6 :** Two spray of Karathane @ 0.05 per cent (Ist spray at initiation of disease and 2nd spray after 15 days)
- **T7:** Control (No spray)

An experiment on chemical management of powdery mildew disease of wheat was conducted at Chatha farm, Jammu during 2009-10 and 2010-11crop season. Powdery mildew susceptible variety, PBW 343 was sown in field in 3 x 4 m^2 plots with 7 treatments including control with four replications in RBD design. Powdery mildew was created by artificial inoculation of spores.

Result

The result revealed that all the treatments recorded significant reduction of disease incidence and increased yield. Two spray of Tilt 25 EC (0.1 per cent) were found highest yield (44.57 q/ha), followed by one spray of Tilt 25 EC (0.1 per cent) at initiation of disease (42.07q/ha) and two spray of Karathane @ 0.05 per cent after 15 days interval (40.47q/ha) respectively. Two spray of Tilt 25 EC (0.1 per cent) were found highest yield increased (10.48 q/ha), followed by one spray of Tilt 25 EC (0.1 per cent) (7.45q/ha) and two spray of Karathane @ 0.05 per cent (6.15q/ha) respectively.

Recommendations

Tilt 25 EC (Propiconazole) (0.01 per cent) at 15 days interval at diseases initiation stage is recommended for complete control of this disease.

Impact

It will enhanced wheat yield 15-25 per cent through disease control and also improved socio economic status of farmers.

Research Stations/Sub Stations: Advanced Centre for Rainfed Agriculture, Rakh Dhiansar, Jammu

56. Integrated Nutrient Management in Rainfed Maize

Contact Scientist: **Dr. Vikas Abrol**

Objectives

To find out the effect of FYM and green manuring practice along with the different dose of nitrogen on the grain yield of maize

Introduction

Dryland soils are generally light in texture, poorly fertile, low in organic matter and exhibit low moisture retention capacity. Maize is the predominant kharif crop and its productivity is adversely affected by frequent dry spells during crop growth period. The technology demonstrated the efficacy of integrated nutrient management by applying FYM @10 t/ha as a source of nitrogen alongwith 40 kg N through urea and recommended dose of phosphorus and potassium for improving maize productivity and soil properties.

Methodology

Well rotten FYM @ 10 ton/ha was incorporated into the soil thoroughly at the time of first ploughing. Maize crop (hybrid) was sown in lines 75 cm apart and the distance of 20 cm was maintained from plant to plant to provide the desired plant population for higher production. Immediately after sowing, atrazine @ 1 kg a.i./ha was sprayed for effective control of weeds.

The entire quantity of inorganic fertilizers *i.e.* recommended DAP and MOP for supplying P (40 kg/ha) and K (20 kg/ha) respectively and 2/3 rd of N (40 kg/ha) through urea were drilled at the time of sowing as a basal dose. Remaining quantity of nitrogen was applied as top dressing in two equal splits - Ist when the crop attained knee high stage *i.e.* after a month after sowing and 2nd before tassel formation *i.e.* about two months after sowing. The plant protection measures were applied as per the package of practice of SKUAST-J

Yield Data with Advantage over Existing Practice

Table 1: Effect of Integrated Nutrient Management on Grain Yield of Maize under Rainfed Conditions

Sl. No.	*Farmer's Name and Village*	*Yield (q/ha)*			*Per cent Increase of Improved Practice over Farmer's Practice*
		Farmer's Practice	*Recommended Practice (100 per cent RDF in maize)*	*Improved Practice (FYM @ 10 t/ha + 40 kg N/ha + Recommended P and K)*	
1	Sukhdev R/o Channi	22.1	26.8	28.3	28
2	Sagar Singh R/o Kartholi	22.2	26.6	25.8	16
3	Khushi Ram R/o Tarore	23.6	29.5	30.1	28
	Mean	22.6	27.6	28.1	24

Per cent increase in mean grain yield was 24 per cent over farmer's practice.

Areas Suitable for the Technology

Rainfed areas of Jammu, Samba and Kathua districts.

Recommendation

Application of integrated use of inorganic and organic fertilizers, *i.e.*, FYM @10 t/ha + 40 kg N/ha (urea) with recommended dosage of phosphorus and potassium resulted in 1.06 times higher grain yield over control (unfertilized) with concomitant 94 and 44 per cent increase respectively in rainwater use efficiency (RWUE) and soil organic carbon over control.

Impact/Benefit of Technology

OFTs were laid on farmers' fields in kharif 2017 and results of the technology demonstrated the efficacy of integrated nutrient management *i.e.* conjunctive use of FYM with inorganic fertilizers for improving the productivity in maize and enhancing organic matter content in resource poor rainfed soils of the region.

57. Biopesticidal Management of Mustard Aphid

***Contact Scientist*: Dr. Reena**

Objectives

1. To assess the insecticidal activity of different extracts against mustard aphid under laboratory conditions
2. To evaluate few best extracts under field conditions

Introduction

Mustard/toria/Gobhi sarson is seriously infested by mustard aphid, inflicting economic losses to the crop. Farmers resort to the application of systemic insecticides like, Imidaclorpid, Thiamethoxam, *etc.* for the management of this pest. This poses serious concerns to the aphid's natural enemies, pollinator bees, besides having several other environmental and ecological concerns. Keeping this in view, experiments were conducted to evaluate the biopesticidal management options against mustard aphid.

Methodology

On Farm trials were conducted during Rabi 2016-17 at three different locations in district Kathua. The experiment comprised three treatments (i) farmer practices (No insecticidal spray) and Improved Practices (*Cannabis* leaf extract and Castor seed extract @ 2.5 per cent). 750 g of cannabis leaves and castor seeds were collected, shade dried, crushed and soaked in 7.5 litre of cow urine for 10 - 15 days. Thereafter it was filtered and the volume was made up to 30 lts with water. This 30 lts of spray solution was sufficient to cover one kanal area of mustard field. As and when the aphid population crossed ETL of 50-60 aphids/10 cm twig, it was sprayed with these solutions. Percent reduction in aphid population in response to sprays after 5 days and seed yields were recorded.

Table 1: Biopesticidal Management of Mustard Aphid in the On-Farm Trials

Location	*No. of Aphids per 10 cm Twig*			*Seed Yield (kg/ha)*		
	Farmer's Practice	*Improved Practice*		*Farmer's Practice*	*Improved Practice*	
		Cannabis Leaf Extract	*Castor Seed Extract*		*Cannabis Leaf Extract*	*Castor Seed Extract*
Rajani Farm (Kathua)	86.2	32.7	26.3	640	820	860
Balbir Dass (Vill. Saida)	90.2	34.5	30.6	620	800	840

Location	*No. of Aphids per 10 cm Twig*			*Seed Yield (kg/ha)*		
	Farmer's Practice	*Improved Practice*		*Farmer's Practice*	*Improved Practice*	
		Cannabis Leaf Extract	*Castor Seed Extract*		*Cannabis Leaf Extract*	*Castor Seed Extract*
Devi Singh (Lacchipur)	77.6	29.1	29.6	640	840	840
Mean	84.67	32.1	28.83	633.3	820	846.7

Per cent reduction in aphid incidence = 62.09 per cent and 65.95 per cent respectively in *Cannabis* leaf extract and castor seed extract

Per cent increase in yield = 29.47 per cent and 33.68 per cent respectively in *Cannabis* leaf extract and castor seed extract

***Cannabis* Leaf Extract**

Castor Seed Extract

Areas Suitable for the Technology

Rainfed areas of Jammu, Samba and Kathua districts

Recommendation

When mustard aphid population crossed ETL (50-60 aphids per 10 cm twig), spraying of 2.5 per cent castor seed extract reduced aphid population by 49.18 per cent and *Cannabis* leaf extract reduced the aphid population by 39.29 per cent in comparison to control and was at par with Thiamethoxam @ 0.33g/lt spray (38.04 per cent reduction in aphid population) after seven days of spray.

Impact/Benefit of Technology

Spraying these biopesticides against mustard aphid not only reduced the aphid infestation, but also conserved the natural enemy (coccinellid and aphid parasitoid, *Diaretiella rapae*) population, which are otherwise killed when the crop is sprayed with insecticides.

58. Mustard Aphid Management through Organic and Inorganic Plant Nutrition

Contact Scientist: **Dr. Reena**

Objectives

1. To assess the impact of various organic and inorganic fertilizers/pesticides on insect pests of mustard.

Introduction

Mustard/toria/Gobhi sarson is seriously infested by mustard aphid, inflicting economic losses to the crop. Crop nutrition is known to affect the insect pests and disease incidence. Experiments were designed to study the impact of organic and inorganic plant nutrition on aphid incidence.

Methodology

Mustard was sown in lines 30 cm apart and the distance of 10 cm was maintained from plant to plant to provide the desired plant population for higher production. The entire quantity of inorganic fertilizers *i.e.* recommended DAP, MOP and urea for supplying P (30 kg/ha) and K (15 kg/ha) and N (50 kg/ha) respectively were applied at the time of sowing as a basal dose. Neem cake @ 5q/ha and MOP @ 40 kg/ha was applied as basal dose. Percent reduction in aphid population and seed yield in response to the treatments (Neem cake and MOP) was recorded.

Yield Data with Advantage over Existing Practice

Table 1: Mustard Aphid Management through Organic and Inorganic Plant Nutrition in On-Farm Trials

Location	*No. of Aphids per 10 cm Twig*			*Seed Yield (kg/ha)*		
	Farmer's Practice	*Improved Practice*		*Farmer's Practice*	*Improved Practice*	
		Neem cake @ 5q/ha	*MOP @ 40kg/ha*		*Neem cake @ 5q/ha*	*MOP @ 40kg/ha*
Rajani Farm (Kathua)	86.2	26.7	23.3	640	860	900
BalbirDass (Vill. Saida)	90.2	23.1	28.6	620	840	820
Devi Singh (Lacchipur)	77.6	24.2	21.9	640	840	880
Mean	**84.67**	**24.67**	**24.6**	**633.3**	**846.7**	**866.7**

Reduction in aphid incidence (per cent) = 70.94 and 70.86 per cent in MOP and Neem cake, resp.

Percent increase in yield = 36.84 and 33.68 per cent in MOP and Neem cake, resp.

Areas Suitable for the Technology

Rainfed areas of Jammu, Samba and Kathua districts

Neem Cake @ 5q/ha

MOP @ 40kg/ha

Areas Suitable for the Technology

Rainfed areas of Jammu, Samba and Kathua districts

Recommendation

Application of MOP (40kg/ha) in addition to the recommended dose or neem cake @ 5q/ha reduced aphid incidence by 41.12 per cent, thereby increasing the grain yield by 18.24 per cent in mustard (*DGS-1)*

Impact/Benefit of Technology

Mustard aphid incidence was greatly reduced in both MOP and Neem cake, thereby enhancing seed yield. Application of neem cake may be utilized as an organic method of managing mustard aphid incidence.

Advanced Centre for Horticulture Research, Udheywalla, Jammu

59. Bio-control of Soil-borne Diseases of Vegetables Crops

***Contact Person:* Dr. Vishal Gupta**

The burgeoning human population in the developing countries like India require increased amounts of food and fiber from a shrinking land base. To date, modern agriculture has largely achieved these objectives by intensive and extensive cultivation of high yielding varieties coupled with chemical fertilizers and pesticides. This chemical pathway to intensify crop production has led to poisoning in human beings and animals and also polluting the environment including air, soil and water resources. Chemical pesticides and their residues have been detected in the food grains, vegetables, fruits, oils, feed, fodder and fiber in most parts of the country. Bio-control agents are a class of environmentally friendly micro-organisms that are increasingly being used to manage the pests including insects, weeds and diseases. *Trichoderma* spp., the beneficial fungi, present naturally in the soil, are known to protect the crops from various soil-borne diseases and also supply nutrients which are essential for crop growth enhancement.

To develop the formaulation of *Trichoderma* spp., local strains were isolated and characterized for their bio-control and plant growth promotion potential by bio-assay followed by pot and field experiments. The methodology for formulation of indigenous and putative strain was standardized and tested against important soil borne diseases such as wilt, collar rot, rhizome rot, corm rot, dry root rot, stem rot and damping off in vegetables crops at multi-locations under farmers' field.

Procedure of Mass Multiplication

The pure culture of *Trichoderma* spp., was inoculated in liquid medium (PD-broth) and kept for 15 days at 26±2°C. The mycelia mat was harvested and mixed with pre-sterilized inert material at 1:3 ratio. After drying for 2 days at room temp., the carboxy methyl cellulose CMC @1 per cent was added having 1x10^{8}/gm cfu.

Formulation of Bio-control Agent.

Demonstration of Application of Bio-control Agent.

Recommendations

Soil Treatment

Mix 1 kg of *Trichoderma* powder in 50 kg of well decomposed FYM in a pit 3' x 6' and cover it with polythene sheet. After every 3-4 days give thorough turning to the FYM and regularly sprinkle small amount of water to maintain humidity. After 15 days the mixture will be ready to apply on one acre of land before sowing.

Nursery Treatment

Bed Treatment

Prepare a suspension by adding 50 g of *Trichoderma* powder in 10 L of water and drench the nursery bed.

Seedling Treatment

Prepare suspension by adding 10 g of *Trichoderma* powder in 10 L of water, dip the seedling roots in it for 30 minutes before transplanting.

Bio-efficiency of *Trichoderma* spp. against Damping Off in Vegetable Crops

Sl.No.	*Treatment*	*Damping Off (per cent)*		
		Tomato	*Brinjal*	*Chilli*
1	Soil treatment with *Trichoderma viride*	38.75	42.75	47.75
2	Soil treatment with *T. harzianum*	32.25	35.25	37.75
3	Soil treatment with *T. harzianum* + *T. viride*	30.67	34.65	35.00
4	Soil treatment with *Carbendazim*	31.55	33.35	34.00
5	Control	55.00	65.00	50.00

Benefit/Advantages

- *Trichoderma* spp. offers effective management against a wide range of soil- borne and seed borne plant diseases.
- Being an excellent saprophyte it has high level of perpetuation in the soil thus increasing its populations and exerting long-term control over phytopathogenic fungi.
- It also helps to decompose organic matter, converting nutrients into available forms so that plants can easily uptake them.
- Due to the indirect effect on crop nutrition it stimulates crop growth because it secretes metabolites that promote the plant development processes.
- It is compatible with most of the commonly used pesticides.
- It is eco-friendly and has important role in sustainable agriculture.
- It is important component of diseases management in organic agriculture.

Organic Farming Research Centre, Chatha

60. NSKE: A Potent Biopesticide for Insect Pest and Diseases Management

Contact Person: **Dr. S.K. Singh**

Introduction

Neem tree (*Azadiracta indica*) is easily available as wild, in the barren lands, around villages and habitats of Sub-tropical zone of Jammu division. This tree requires very less nutrition, care and provides good soil cover against erosion but it has not been exploited for its pharmacological properties. Its fallen leaves, bark and seed kernels possess high pest-control ability and can be used as a potent bio-pesticide against potential insect- pests of agricultural crops. It can help to generate revenue for the poor and unemployed youth, who can collect and efficiently utilized them for plant protection as Neem Seed Kernel Extracts (NSKE.). Village-level industries can be set up for the production of different products from this tree, which otherwise goes waste. In sub-tropical zone of Jammu division, no neem garden/orchard has been developed for this purpose, thereby opening an option for the scientific cultivation of neem. NSKE would be environmentally non pollutive, renewable, inexhaustible, indigenously available, easily accessible, largely non-phyto-toxic, systemic, easily biodegradable and relatively cost effective.

Preparation of NSKE

1. Collect ripened neem seeds.
2. Clean them and remove their outer shell.
3. 12-15 kg (fresh seeds) or 18.0 kg (old seeds) are required for the preparation to NSKE/hectare.
4. Crush the kernels gently and kept them loose with a cotton cloth bag.
5. Left for soak them overnight in a vessel @ 50 g/L.
6. Filter it with muslin cloth.
7. On filtering, 500-600ml of extract is obtained.
8. Diluted 1.0 liter of this extract with 20.0 liters of water to get 5 per cent of the solution.
9. This concentration of the extract can be increased or decreased depending on the intensity of pest attack.

Mode of Action

Neem seed contains 0.1-0.3 per cent Azadirachtin-an active ingredient, which disrupts the development of the insect and deters their feeding. It also acts as antifungal agent, releasing different alkaloids, which harms or kills the pathogen. NSKE posses multiple mode of action against insect and diseases, whereas not harmful to birds and mammals.

Insect/Pest Management: Aphid Management in *Brassicas*

Aphid attack in mustard was managed by spraying 5 per cent NSKE by reducing its population more than 75.0 per cent. Aphid counts were calculated from the top 10 cm central infested twigs, before and after spraying of NSKE.

Mustard Crop before the Spray

Mustard Crop after the Spray

Reduction in aphid population in crops like mustard, broccoli, cauliflower, radish, turnip and knolkhol were more than 85 per cent after 7th day spray of spray with mean value of more than 40.0 per cent in all the crops (Table 1) particularly in the plots having vermi-compost @2.5 t/ha.

Table 1: Effect of Foliar Apray of 5 per cent NSKE on Aphid Management in different *Brassicas* during Kharif 2016 and 2017 at OFRC, Chatha

Treatments	*Percent Reduction in Aphid Population*			
	*3rd DAS**	*5th DAS*	*7th DAS*	*Mean*
Mustard	7.60 (12.5)	44.0 (35.90)	78.5 (61.88)	43.37
Broccoli	6.28 (14.50)	45.10 (42.16)	87.14 (69.53)	46.17
Cauliflower	8.23 (16.66)	47.35 (43.46)	88.98 (70.81)	48.19
Radish	5.30 (13.29)	44.88 (42.04)	87.14 (69.25)	45.77
Turnip	5.61 (13.69)	42.18 (40.48)	85.38 (67.62)	44.39
Knol khol	5.41 (13.44)	31.88 (34.36)	86.38 (68.84)	41.22
C.D (0.05)	0.96	3.38	N.S.	-
S EM+	0.29	1.44	2.96	-
C.V.	3.51	4.36	7.42	-

*DAS = Days after spray.

Figures in parenthesis are arc sine transformed value

Disease Management

Brown Leaf Spot Management in Rice

Seedling dip treatment- Dipping of rice seedlings for 30 minutes in 5 per cent NSKE solution, effectively reduced brown leaf spot disease upto 37.17 per cent. These seedlings were transplanted in already prepared/amended soil with *Trichoderma harzianum+ Pseudomonas fluorescens* @5kg/ha (1:1) in all treatments except control or unamended plot (Table 2)

Rice Crop before the Spray

Rice Crop after the Spray

Table 2: Effect of 5 per cent NSKE as Seedling Dip Treatment for Management of Brown Leaf Spot Disease of Rice during Kharif 2016 and 2017 at OFRC, Chatha

Sl.No.	Treatments	Disease incidence (per cent)
1	*Trichoderma harzianum*@ 5g/L of water	16.53 (23.93)
2	*Pseudomonas fluorescens* @ 5g/L of water	17.88 (24.97)
3	*Trichoderma harzianum*+ *Pseudomonas fluorescens* @ 5g/L of water (1:1)	14.79 (22.58)
4	Organic Zyme (Phosphorus rich PGPR) @5g/L of water	18.97 (26.46)
5	*Glomus mosseae*@5g/L of water	20.87 (27.16)
6	NSK.E@ 5 per cent	14.59 (22.43)
7	*Trichoderma viride* @5g/L of water	19.43 (26.10)
8	Control	23.22 (28.79)
	C.D (0.05)	2.33
	S EM+	0.76
	C.V.	5.22

Figures in parenthesis are arc sine transformed values

Benefits/Impact

NSKE @ 5 per cent foliar spray in *Brassicas* and seedling dip treatment in rice resulted in input cost reduction by 1.5 times as compared to chemical treatment by synthetic pesticides (Table 3).

Table 3: Economic Comparison between NSKE and Synthetic Pesticides

Items	NSKE	Synthetic Pesticides
Quantity/ha	12-15 kg neem seed kernels	Imidacloprid (Single spray) 200 to 300 ml
Cost of product/ha (Rs)	240-300 *i.e.*, 20/kg	Rs 300-450 @ Rs 150/100ml
Other costs (Rs.) (Grinding and extraction)	24-30	-
Spray cost/ha (Rate/manday)	324	324
Total cost/ha	588-654	624-774

Recommendation

I. Neem Seed Kernel Extract (NSKE) can be used either as foliar spray or seedling dip treatment against aphids in crucifers and brown leaf spot disease in rice.

II. It must be applied at the rate of 5 percent as prophylactic measure and after first appearance of insect/pest and disease.

III. It is recommended that its spray should be repeated at fortnightly intervals to get better results.

IV. It is very important to amend the soil with *Trichoderma harzianum*+ *Pseudomonas fluorescens* @5kg/ha (1:1) in rice and Vermicompost @2.5t/ha in *Brassicas* for attaining good results.

61. High Density Organic Guava Cultivation

Contact Scientist: Dr. Akas Sharma

Guava (*Psidium guajava* L.) known as poor mans' apple is one of the important fruit crop grown in the sub-tropical areas of Jammu Province. With the growing awareness for safe and healthy food among consumers and growing concern for deteriorating soil health, fertility and depleting natural resources among policy planners, strategies are being drawn to adopt technologies promising safe and healthy food with due respect to the natural resource availability without compromising the productivity and food security. In these concerns, organic guava cultivation will emerge as one of the viable option which promises food production in quantity and quality with resource sustainability under Jammu Sub-tropics.

Methodologies

The investigations on "Effect of FYM and vermicompost on high density guava orchard for its early establishment under organic management system" was carried out at Organic Farming Research Centre, SKUAST-J, Chatha, Jammu. Initial Soil Analysis of the orchard area showed that organic carbon was in the medium range. As per the technical programme the high density orchard guava cv. Sardar (L-49) was planted during March, 2016 at a spacing of 6.0metre x 3.0metre (rows x plants). Planting of orchard was done on the raised beds of 5.0 feet from the ground level which was made with the help of JCB. Before planting, layout was done for pit digging, pits of size 0.5 m x 0.5 m x 0.5 m (LxBxD) were dugged out and the dugged soil was mixed with well rotten FYM@40 kg and neem leaves @20kg. Monitoring and management of pest and diseases in the guava orchard was done organically. The application of fertilizer was done as per the technical programme during the first week of July.

The pruning of plants was done in the month of April, 2016 soon after one month of planting by cutting the top of the plant upto the height of 60-70cm from the ground level which continues with three prunings in the initial years of establishment*i.e.* First- pruning of newly emerged shoots from the top cut and retaining only 3 to 4 shoots, Second - six months after planting by cutting back the shoots to 50 per cent of their total length and Third-pruning of downwards/criss-cross and shoot tip pruning after twelve months of planting. After pruning, cuts were treated with fresh cow dung to avoid any fungal infection. Fresh cow dung releases anti-pathogenic enzymes which lyses the cell wall of the fungus and also suck the protoplast, ultimately killing the pathogen. Flowering was regulated for winter season crop *i.e.* MrigBahar which was started in the month of August. Harvesting of fruits was done during December-January. The yield and the operating cost of production were worked out for the period orchard establishment.

Rodent Management

Rats are major pests in orchard, they makes holes around plants and its periphery and damaging root system. The major practices were applied to manage rodent population-

1. Fixing of Rattol Glue Trap @ 4/Kanal area, to attract and trap the rat.
2. Catching two rats and painting with affraiding colour to quarrel among the rodents
3. Stacking of the bamboos in the orchard field for sitting of the birds in night as well as days for hunting rats. Owl is the best predator which hunts rodents at night.

Cost of Establishment

During the first year of planting the organic guava orchard with spacing 6.0 m×3.0m, major expenditure was incurred on the preparation of land, hired labour, planting material, digging, filling of pits, planting and farm yard manure, *etc.* The operation wise per hectare establishment cost for the orchard is presented in Table 1. The per hectare total costs incurred during first year for the establishment of guava orchard were found to be Rs. 277140.71. Variable costs per hectare were found to be Rs. 237685.00 with Rs. 100000.00 incurred on preparation of land. Expenditure for preparation of land was highest because of Rs. 90000.00 incurred on hiring of JCB for levelling of land. It was further observed from the table that Rs. 45000.00 was incurred on hired labour, Rs. 27750.00 on digging of pits, Rs. 13875.00 on filling of pits and planting, Rs. 24975.00 on planting material, Rs. 17760.00 on farm yard manure and minimum of Rs. 8325.00 on neem cake. The per hectare interest on working capital was worked out to be Rs.14261.10 whereas, Rs. 25194.61 was incurred as management cost.

Table 1: Operation-wise First Year Establishment Costs of Guava Orchards with Spacing 6.0 m x 3.0 m (Rs./ha)

No. of Plants per hectare = 555

Sl.No.	*Item*	*Cost (in Rs.)*
1	Preparation of land (i)+(ii)	100000.00
	i. Tractorization for levelling of raised beds under high water table	10000.00
	ii. JCB (100 hrs @Rs 900/hr)	90000.00
2	Hired Labour @Rs 150/manday for 300 mandays	45000.00
3	Digging of pits @Rs 50/pit	27750.00
4	Filling of pits and planting @Rs 25/pit	13875.00
5	Cost of Planting material @Rs 45/plant	24975.00
6	Farm Yard Manure (40kg/pit) @Rs 1200 for 15 quintals	17760.00

Sl.No.	Item	Cost (in Rs.)
7	Neem Cake (½kg/pit) @Rs 30/kg	8325.00
8	Total Variable cost (1 – 7)	237685.00
9	Interest on working capital @6 per cent	14261.10
	A. Total cost of cultivation (8+9)	251946.10
	B. Management cost @10 per cent of cost of cultivation	25194.61
10	Establishment cost (A+B)	277140.71

Cost and Return Analysis for Initial Phase of High Density Guava Plantation

The cost and return analysis determines the economic feasibility of high density plantings. The net income was worked out as a difference between gross income and cost of cultivation at respective costs during first fruiting crops depicted in Table 2. Major expenditure of Rs.30000.00 during first year of fruiting was incurred on hired labour @Rs 150/manday for 200 man-days. The net income per hectare is ranged from minimum of Rs. 4480.00 in treatment T_9 and maximum of Rs. 36498.00 in treatment T_8. It is further observed from the table that treatments (*i.e.* T_1, T_2, T_3, T_4, T_6, T_7, T_8) were in profit indicating positive net income while rest treatments resulted in negative net income. Particularly only two treatments *i.e.* T_7 and T_8 recorded significant profit more than Rs. 35000.00 during first year of fruiting, amongst which treatment T_8 was considered as best treatment with the maximum profit of Rs. 36498.00 and benefit cost ratio of 1:2.21.

Table 2: Yield of High Density Guava cv. (L-49) Orchard, Profit and Loss Account B : C Ratio under Organic Orchard Management

Treatments	Yield Quintal/ ha	Gross Income/ ha (Rs)	Cost of Cultivation/ ha (Rs.)	Net Income (Rs.)	B : C Ratio (t)
T_1 - 20 kg FYM	09.71	38840.00	30016.00	8824.00	1:1.29
T_2 - 20 kg FYM + organic mulching	10.82	43280.00	30016.00	13264.00	1:1.44
T_3 - 25 kg FYM	09.71	38840.00	30020.00	8820.00	1:1.29
T_4 - 25 kg FYM + organic mulching	11.65	46600.00	30020.00	16580.00	1:1.55
T_5 - 15 kg vermicompost	06.94	27760.00	30135.00	2375.00	1:0.92
T_6 - 15 kg vermicompost + organic mulching	10.26	41040.00	30135.00	10905.00	1:1.36
T_7 - 15 kg FYM + 10 kg vermicompost	16.30	65200.00	30102.00	35098.00	1:2.17

Treatments	Yield Quintal/ ha	Gross Income/ ha (Rs)	Cost of Cultivation/ ha (Rs.)	Net Income (Rs.)	B : C Ratio (t)
T_8 - 15 kg FYM + 10 kg vermicompost + organic mulching	16.65	66600.00	30102.00	36498.00	1:2.21
T_9 - Absolute control(No fertilizer)	06.38	25520.00	30000.00	4480.00	1:0.85

Physico-chemical Characteristics of Organic Guava Fruit

The physico-chemical analysis of organic guava showed that maximum fruit length x breadth (8.56 x 8.06 cm), weight (196.46 g), pulp weight (163.31g) and fruit set (79.31 per cent) respectively were recorded with the treatment comprising 15 kg FYM + 10 Kg vermicompost + organic mulching applied to the one year old guava plants similarly, the highest yield (16.35 quintals ha^{-1}) for winter season crop with TSS (13.26oB), total sugars (8.73 per cent), Vitamin 'C' (217.83 mg per 100 g of pulp) and pectin (0.77 per cent) respectively were recorded with the same treatment.

Planting of High Density Organic Guava Orchard at OFRC, SKUAST-Jammu

Steps to be involved for development of training system inhigh density Organic guava orchard

Layout for Planting at 6.0 x 3.0 metre Spacing

Planting of Guava cv. Lucknow-49

Irrigation after Planting

Irrigation with Half Moon System

Selection of Middle Shoot for Top Cut

Pruning of Shoots to Single Stem

Pruning of Side Shoots

Measuring the Plant Height for Top Cut

Cutting the Top of the Plant to 60 -70cm for 6.0 x3.0 metre Spacing

After Pruning of Guava Plant

High Density Organic Guava Orchard at Organic Farming Research Centre (OFRC) SKUAST-J, Chatha, Jammu

Organic Guava Orchard after Six Months of Planting at OFRC, SKUAST-J, Chatha-Jammu.

Organic Guava Orchard after One Year of Planting at OFRC, SKUAST-J, Chatha-Jammu.

Established High Density Organic Guava Orchard with Uniform Canopy Architecture.

Pasting of Fresh Cow Dung to Manage Fungal Infection after Training and Pruning of Guava Plants.

White Mycelial Growth on Trunk (un-pasted cow dung or control).

Yearly Pruning of Guava Trees under High Density

Pruning is the most important operation in guava cultivation, because blooming always comes out from the new branch. A good tree-form (tree shape) is required to maintain the productivity of guava trees. Guava trees should be dwarfed for better field management. There are three different types of pruning: heavy pruning, medium pruning and regular pruning.

Regular pruning keeps the sub-main branches. This type of pruning is always used in year-round production of fruit

- Prune the downward growing branches as they do not produce good fruit.
- Sucker branches grow erect and do not bear fruits – prune
- Overlapping branches need to be cut off (pruned) to allow more sunlight through the branches.
- High branches are cut off also.
- Diseased branches are cut off to eliminate a potential source of disease.

Flower and Fruit Thinning

Flower and fruit thinning is always done to control fruit quality and quantity. In general, flower buds appear on the 2ndto 5thpair of leaf axils. The flowers are usually borne in single, double and triple cluster. The fruiting branch should be controlled by pinching (about 1-1.5 ft.) to reduce excessive flowering and poor fruit growth. In double type flowers, one flower is kept; usually the flower which is not too exposed to sunlight and in the case of triple flowers, only the center flower is left. After flower thinning, there are always about 2-4 flowers on one fruiting branch and when the fruit sets; fruit thinning should be done. For good fruit characteristics and quality, it is recommended to keep 1-2 fruit on one fruiting branch.

Calendar of Operations for Establishing High Density Plantation of Organic Guava Orchard at Spacings of 6.0 x 3.0m (Rows x Plants)

May

- Ensure the soil suitability by digging a profile pit of 90 cm depth.
- Observe the soil for presence of any calcareous nodules or sheet rock. If the lime nodules or sheet rock is present avoid such soils for taking up plantations.
- Based on soil test report, even if the soil pH is up to 8.2 it is considered as suitable for guava plantation.

June

- Take up summer ploughing with MB plough twice and level the land.
- During layout, care should be taken that row to row spacing should be facing east-west direction.
- Alignment and peg marking for pit digging.
- Pit size to be followed is 0.5x 0.5x 0.5metre.

- Allow the pit for exposure to sun at least for two weeks before filling the pits to control soil born pest and diseases.
- Make arrangements for procurement of plant material from reliable nurseries preferably from Govt. nurseries or from nurseries recommended by National Horticulture Board.
- High yielding varieties like Lucknow - 49 (Sardar guava)/Shweta may be selected.

July

- Pit filling – fill the pits firstly with 40 kg decomposed FYm along with 20 kg of neem leaves
- Fill the pits with the soil at least 6" above the ground level so that when the soil settles it will be at the ground level at planting.
- Ensure thorn fencing or with any thick growing live fence to control cattle trespassing.

August

- Budded/layered guava plants should be planted by keeping the plants in the centre of the pit without disturbing the ball of earth around the root system.
- Water the plant immediately after planting.
- Irrigate the plants at every 4 to 5 days interval during the first month and later at 7 to 10 days interval if there are no rains during the initial two years.

September

- Staking of the plants should be done for its upright growth.
- Weeding should be done at least in a fortnight during monsoon season to control weeds. Simultaneously the area around the plant should be made to retain water during dry spells.
- To avoid Evapo-transpiration losses, mulch the basins with paddy husk or paddy straw or ground nut shells. Mulching also help to control the weeds in the basins.
- If heavy rains persist leading to ill-drained conditions, provide drainage channels to avoid water stagnation. Ill-drained condition leads to yellowing of leaves and wilting of plants.

October

- Trees planted during July –August should be topped to a uniform height of 60-70cmfrom the ground level to induce the emergence of new growth below the cut point.
- Three to four equally spaced shoots should be retained around the stem

to form the main scaffold limbs of the tree and allowed to grow for 4-5 months till they are 40-50 cm in length.

- Work out the soils in the basin to keep weed free and improve aeration.

November

- Prune the side shoots which appear from time to time to 50 per cent of their length to encourage lateral branches preferably in the east -west direction for better sunlight exposure.
- If inter crop during June-July is not cultivated, farmers can sow inter crops for rabi season during this month.
- For the management of mealy bug, the soil under the trees should be dug up after the monsoons and treated with 10 per cent cow urine.

December

- Newly emerged shoots are allowed to grow up to 40 -50 cm and are pruned again to 50 per cent for emergence of new shoots.
- Ensure regular irrigations at 7-10 days interval.
- To protect them from mealy bugs, Adhesive bands (prepared by boiling together rosin 6lb and castor oil 5 lb for 10 minutes) should be placed around the trunks of the trees in December.
- If white fly incidence is noticed on matured leaves curling and reddening of leaves will be observed. Control by spraying 5ml of neem oil in 1lt of water.

January

- Ensure regular irrigations at 7-10 days interval.
- Reddening of leaves due to phosphorous deficiency may be observed. It may occur due to immobilization of phosphorus due to low soil temperatures in winter season. Correct it by 1 kg rock phosphate per tree per year applied in split doses once in Jan and August and frequently irrigating in the basins.

February

- Gap filling - If any gaps are noticed due to death of plants, it should be replaced by taking up gap filling.
- Weeding and soil working to improve aeration and weed control

March

- For 6 x 3 metre spacing, all the plants are confined to a hedge shape of 2 m inter row width and 5 metre height for which pruning is performed in February and shoot tip pruning (10-15cm tip) July-August every year.
- Remove the side shoots/stock sprouts which appear from the base of the plant.

- Irrigate the plants at 5-7 days interval regularly.

April

- Ensure regular irrigation through drip or at least once in 5-7 days interval through basins irrigation.
- The tender shoots will be infected with mealy bugs or white flies. Control them by spraying Neem Oil (0.5 to1 per cent).
- White Fly incidence will also be severe during this month. Trap the white flies by keeping yellow cards or metal boxes applied with grease or any sticky substance.

Flow Chart for Establishment of High Density Organic Guava Orchards

Spacing 6.0mx3.0m

(Row to Row x Plant to Plant)
Accommodates 555 plants ha^{-1}

↓

Pit size =0.5 x 0.5 x 0.5 metre (LxBxH)

↓

Planting in Feb-March/July-August

↓

Cut the top of the tree at height of 60-70cm from the ground level after 2-3 months of planting

↓

Pasting of fresh cow dung on cut endsto avoid fungal infection

↓

New shoots emerge below the cut point
Retain 3 to 4 shoots (Equally spaced)

↓

Prune the shoots after 3-4 months of shoot emergence (cutting back to 50 per cent of their total length)

↓

New shoots emerge below the cut end

Further prune the shoots after2-3 months of emergence
(cutting back to 50 per cent of their total length)

↓

Continue shoot pruning during the second year for desired tree shape

↓

After second year

↓

Shoot pruning
February every year
Plant height confined upto 5.0 m

↓

Continue shoot pruning (50 per cent) on the tree every year
in the month of February and shoot tip pruning in
July-August to maintain the tree shape and size

Recommendations

1. Application of Farmyard manure and Neem Cake @ 2:1 ratio (i.e 40kg FYM and 20kg Neem cake) in the pits before planting encourages the cent percent survival of guava plants.
2. Application of fresh cowdung on the cut ends of the pruned shoots protect the guava plants from fungal infection
3. Before and after rainy season one kg of decomposed FYM enrich with *Trichoderma spp.* applied to the basins of each guava plants found to be effective for reducing the incidence of soil borne diseases.
4. It is recommended economically also, as it will enhance the farmers' income with a profit of Rs.36498 and benefit cost ratio of 1:2.21 during the first year of production.

Therefore, high density organic guava (cv. L-49) orchard standardized at a spacing of 6.0 x 3.0 (rows x plants) is the best for commercial cultivation under Jammu sub-tropics.

Benefit/Impact

Establishing an Organic guava orchard will be safe and eco-friendly in nature. It will reduce the usage of chemical and pesticides which are the main causes for deteriorating the soil as well as human health. Orchardists can use locally available manures for orchard fertilization and can biologically control pest and disease management while, establishing high density organic guava orchard. It will secure the socio-economic conditions of the orchardists as guava is one of the cash crops of this region, it will definitely enhance the farmers' income.

RHRSS, Bhaderwah

62. Effect of Nutrient Management in Maize + Rajmash Intercropping in Sub-alpine Conditions of Bhaderwah

Contact Scientist: **Dr. Rohit Sharma**

Methodologies

A field experiment was conducted on, "Effect of nutrient management in maize+rajmash intercropping in sub-alpine conditions of Bhaderwah" during kharif 2011 and 2012 and subsequently the recommendations has been deliberated in the RCM further the recommendations were validated in farmers field by conducting On Farm trial during Kharif season 2016 and 2017 at two sites of Bhadarwah *i.e.* sartangal (1880 amsl) and Gwari(1710amsl) for technology generation. The detailed technical programme and the agronomic management practices of the same in OFTs trials is as under

A. Treatment Details of OFTs

Treatments	*Area*	*No. of OFTs*
$T_{1:}$New recommendations (one row of maize followed by one row of rajmash with fertilizer dose of 75 kg/ha urea, 68 kg/ha DAP and 25 kg/ha MOP and 3 t/ha FYM)	0.5 Kanal	05
$T_{2:}$ Existing recommendations ***i.e.*** **Package of practices :** (Mixed cropping using seed rate of 8:1, with fertilizer dose 100 kg/ha urea, 90 kg/ha DAP and 33 kg/ha MOP)	0.5 Kanal	05

B. Agronomic Management Practices for New Recommendations (T_1) and Existing Recommendations (T_2) for OFTs Trials

Sl. No.	*Management Practices*	*New Recommendations (T_1) (NR)*		*Existing Recommendations (T_2) (ER)*	
		Maize	*Rajmash*	*Maize*	*Rajmash*
1	Date of sowing Sartangal Gwari	 29th April 10th May	Same as for maize	 29th April 10th May	Same as for maize
2	Variety	Bioseed 9621	Local (Chinta selection)	Bioseed 9621	Local (Chinta selection)
3	Seed rate (Kg/ha)	40	24	40	5.0
4	Spacing	75X20	One row of maize is followed by one row of rajmash	Mixed cropping using seed rate 8:1	Mixed cropping
5	Fertiliser dose (Kg/ha) Urea DAP MOP FYM	Total Basal top dressing 1st 2nd 75 33 21 21 68 68 - - 25 25 - - 3 3 - -	Fertiliser requirement as per the base crop	Total Basal top dressing 1st 2nd 100 56 22 22 90 - - 33 - -	Fertiliser requirement as per the base crop
6	Fertiliser application	1st top dressing at Knee high and 2nd top dressing at Tassel stage		1st top dressing at Knee high and 2nd top dressing at Tassel stage	
7	Weed control	Hand weeding and hoeing 25 DAS Knee high stage	No additional requirement	Two hoeing 15 DAS 30 DAS	No additional requirement
8	Date of harvesting	1st fortnight of October	Pods are hand picked in intervals when they mature	1st fortnight of October	Pods are hand picked in intervals when they mature

Table 1: Effect of Existing Recommendations and New Recommendations on MEY (q/ha), Net Returns and B : C Ratio of Maize + Rajmash Cropping System (Average of two years)

Sites	*Maize Equivalent Yield(q/ha)*		*Percentage Increase in MEY by New Recommendations over Existing Recommendations*	*Net Returns (Rs)*		*Increase in Net Returns by New Recommendations over Existing Recommendations (Rs)*	*B : C Ratio*	
	ER	NR	39.29	ER	NR	36896.17	ER	NR
Site-1 Sartang at (Mean of three OFTs)	76.99	107.24		82949.33	119845.5		2.54	2.92
Site-2 Gwari (mean of two OFTs)	59.47	92.34	55.27	56667.75	97467.75	40800	1.75	2.38

Table 1 revealed that the highest average MEY of 107.24 q/ha and 92.34 q/ha was recorded in new recommendations in both the sites of Sartangal and Gwari respectively. The net returns(Rs 119845.5 in Sartangal and 97467.75 in Gwari) and B:C ratio(2.92 in Sartangal and 2.38 in Gwari) was also found highest with new recommendations.

Sowing of Crop on Farmer's Field.

Agronomic Management Practices on Farmer's Field.

Crop under New Recommendations.

Crop under Existing Recommendations.

Recommendations

Hence it is recommended that in hilly tract of Jammu the new recommendations where one row of maize is followed by one row of rajmash (1:1) with fertilizer dose of 75kg urea/ha,68kg DAP/ha and 25 kg MOP/ha and 3t FYM/ha produced the highest MEY, Net reurns and B:C ratio. Thus New recommendations is the best option to enhance productivity of maize under intercropping with rajmash as compare to existing recommendations or package of practices.

Benefits/impact

1. Unfortunately in Bhadharwah there is, at present, no proper and economically viable inter-cropping system in practice. The farmers sow maize with rajmash by just mixing the seed without any proportion and row arrangement. Thus, it is imperative to develop efficient system of maize + rajmash cultivation to increase productivity per unit area and time.
2. As the new recommendations enhanced the productivity of maize by increasing MEY to 39.29 and 55.27 per cent over existing recommendations in Sartangal and Gwari,respectively. Also the net returs increased to the tune of Rs 36896.17 and Rs 40800 over existing recommendations in both the sites *i.e.* Sartangal and Gwari,respectively. Thus by using new recommendations the viability and sustainability of intercropping system increased on sustained basis besides conservation of resource base in undulated topography in hilly regions can also be ensured.

63. Agrotechnique Refinement for Wild Pomegranate (*Anardana*) in J&K

***Contact Scientist:* Dr. P. Bakshi**

Introduction

Wild pomegranate, *Punica protoprunica* is one of the minor fruits of great economic importance in the vast tract of the hill slopes of Jammu and Kashmir. It grows in wild state as a large evergreen shrub, 4 to 6 m high and, flowers during the months of May and June. The fruits ripen towards the middle of October and are hand picked. This fruit is filled with angular hard seeds covered with a juicy, pink or yellowish white sweet astringent acid pulp. The seeds along with the juice sac are separated from the rind and spread in the fields, rooftops or over stone slabs in open sun till dry enough for safe storage. The drying process normally takes 10 to 15 days when the colour turns reddish brown. The sun-dried seeds with pulp constitute the condiment '*Anardana*' which is extensively used as an acidulant in most of Indian curries. It is also used in ayurvedic medicines, tanning and colouring industries. Every year, *Anardana*, worth crores of rupees is collected from the hills and sold at various places throughout the country and abroad. Besides, *Anardana*, huge quantities of the fruit rind, worth thousands of rupees is also exported for use in various industries.

The obvious reason for its low production in J&K is attributed to the severe infestation of anar butterfly coupled with lack of suitable agro-techniques for qualitative and quantitative production. The demand of *Anardana* has gone up recently due to the opening of entirely a new market for the commodity in the West Asian countries but there is wide variation in produce making this industry less profitable. So, it is imperative for the researchers to divert their attention towards growing of wild pomegranate on scientific lines and making it more remunerative so that the variability on the product is reduced and high quality *Anardana* can be obtained which can be exported to other countries also. Before the wild pomegranate is propagated vegetatively and plantation programme on mass scale is taken up, it is the need of hour to select a suitable germplasm for high qualitative and quantitative traits. Further at present the *Anardana* industry in our country in general and J&K state in particular is encountered with some problems like fruit cracking and ravages of pomegranate butterfly in Jammu and Kashmir. Keeping in view the above mentioned threats, the present study was conducted to select the improved type of strains available in Udhampur, Doda and Rajouri districts for commercial cultivation in the suitable regions. Thus, this research proposal has been formulated with the following objectives:

Objectives

1. Survey, selection, evaluation and multiplication of wild pomegranate germplasm of Jammu province.
2. To assess the prevalence and role of promising natural enemies in wild pomegranate ecosystem and to standardize the mass production techniques for suitable strains.
3. To study the incidence and control of fruit cracking in wild pomegranate and
4. To conduct field demonstration on field effectiveness of bio-agents in selected locations

Methodology

Survey, Selection and Evaluation of Wild Pomegranate Germplasm for Superior Traits (1st Year)

A survey work was carried out in all the three *Anardana* growing districts namely Udhampur, Doda and Rajouri. After consultation with C.H.O and H.D.O's, the blocks which have highest production and maximum area under *Anardana* were selected from Udhampur, Doda and Rajouri. From Udhampur, Chennani block; from Doda, Assar block and from Rajouri, Manjakota block has been marked out. The areas of Chennani are: Champairi, Dogra Gavoh and Sudhmahadev; of Assar are Assar, Baghar, Kaurapani, Malhori and Khellani and from Manjakota are Badal, Gadyog, Jamora, Charhan, Darsal, Gambhir, Patrara, Panjgrain, Saroola, Swari and Tralla. On the basis of various growth parameters as well as physico-chemical characteristics of fruits, the plants from Champairi, Chennani and Dogra Gavoh of Udhampur; Assar, Malhori, Khellani and Kaura pani of Doda and Badal, Tralla, Swari, Gambhir and Patrara areas of Rajouri districts were found suitable for multiplication. The hardwood cuttings (2600 nos.) from the selected areas have been planted at Research farms of RHRSS, Bhaderwah and Division of Fruit Science, Udheywalla, respectively during 2006-07. The plants raised from these cuttings shall be distributed among the farmers in the ensuing year.

Results

a. Physico-chemical Characteristics

Table 1: Physical Characteristics of *Anardana*

Sl. No.	*Area*	*Fruit Weight (g)*	*Fruit Length (cm)*	*Fruit Breadth (cm)*	*Fruit Colour (Visual)*	*Fresh Aril Colour*	*Dried Aril Colour*
1.	**Udhampur**	172.30	8.01	6.95	Brown, light brown and dark red	Red	Red

Sl. No.	Area	Fruit Weight (g)	Fruit Length (cm)	Fruit Breadth (cm)	Fruit Colour (Visual)	Fresh Aril Colour	Dried Aril Colour
2.	**Rajouri**	148.80	7.37	6.02	Light brown and red	Red	Red
3.	**Doda**	131.60	7.42	5.97	Light brown, red and dark red	Red	Red

Table 2: Chemical Characteristics of *Anardana*

Sl. No.	Area	TSS (° Brix)	Acidity (per cent)	Reducing Sugar (per cent)	Total Sugar (per cent)	Sugar : Acid Ratio	Tannins (mg/100 ml)
1.	**Udhampur**	16.0	7.20	13.1	14.3	1.99	156.2
2.	**Rajouri**	16.0	7.10	12.6	14.8	2.09	148.6
3.	**Doda**	15.5	6.80	12.2	15.8	2.32	149.7

Table 3: Processing Characteristics of *Anardana*

Sl. No.	Area	Juice (per cent)	Seed (per cent)	Rind Thickness (mm)	Dehydration Ratio	Rehydration Ratio
1.	**Udhampur**	51.6	51.63	4.25	3.98:1	1:2
2.	**Rajouri**	49.6	51.00	3.86	3.96:1	1:2
3.	**Doda**	49.8	49.86	3.69	4.01:1	1:2

These are the mean values of all the characteristics as there were three replications per district with 2.5 kg fruit per replication for chemical and processing characteristics, while for physical characteristics, 10 randomly selected fruits were taken for each replication.

b. Different Types of Drying

Four different types of drying was done after separating the seeds from rind and septa, so as to find out the best method of drying for anardana.

- Vacuum drying
- Oven drying
- Sun drying
  - Ordinary
  - Poly tent house
- Ambient temp.

Table 4: Effect of different Types of Drying on Quality of *Anardana*

Sl. No.	*Drying Method*	*Duration*	*Moisture (per cent)*	*Dehy-dration Ratio*	*Organoleptic Rating*			*Overall Acceptance (15.0)*
					Texture (5.0)	*Taste (5.0)*	*Colour (5.0)*	
1.	**Vacuum**	13 hrs (42 + 2 °C)	73.88	3.83:1	3.5	3.0	3.0	9.5
2.	**Oven**	16.5 hrs (42 + 2 °C)	75.62	4.02:1	3.0	3.0	3.0	9.0
3.	**Sun**	38 hrs	73.62	3.81:1	4.0	4.5	4.5	13.0
4.	**Poly-tent house**	42 hrs	71.27	3.48:1	4.5	4.0	4.0	12.5
5.	**Room**	10-12 days	75.12	4.13:1	2.0	3.0	2.0	7.0

Out of four different types of drying, sun drying for 38 hrs was found to be best followed by poly-tent house sun drying. However, the seeds dried at room temp. showed some fungal growth. There is no microbial growth in all other types of drying.

c. Propagation of *Anardana*

The propagation of wild pomegranate by seed is easy and seedling plantations appear hardier, drought resistant due to deep tap root system but seed propagation brings genetic variability and leads to low yield and poor quality of fruits. The vegetative method of propagation is advocated for eliminating high degree of variability observed in the seedling plantations. Propagation through hard wood cutting is best and less expensive method of its propagation. The cuttings taken from Assar, Baghar and Kaurapani areas of Doda district were raised at RHRSS, Bhaderwah and FoA Udheywalla. The plants raised at RHRSS, Bhaderwah were handed over to Incharge, RHRSS Bhaderwah for their plantation at boundary wall, while those raised at FoA Udheywalla were planted at boundary of Nursery of Fruit Science Section Chatha in February, 2008.

Table 5: Propagation of *Anardana*

Accessions	*Survival (per cent)*	*Height (cm)*	*No. of Shoots*	*No. of Primary Roots/Plant*
WP-1	29.26	57.86	3.92	4.02
WP-2	27.30	56.28	3.95	3.94
WP-3	30.02	58.12	4.02	4.26
WP-4	26.82	54.26	3.72	3.76
WP-5	25.04	48.26	2.91	2.92

Accessions	*Survival (per cent)*	*Height (cm)*	*No. of Shoots*	*No. of Primary Roots/Plant*
WP-6	27.01	56.43	3.84	3.84
WP-7	25.46	49.82	3.66	2.86
WP-8	26.84	51.68	3.74	3.41
WP-9	26.37	52.14	3.89	3.32
WP-10	25.92	49.48	3.72	3.08
WP-11	26.64	55.21	3.86	3.62
WP-12	27.16	56.34	3.94	3.86
C.D. 0.05	1.15	1.18	0.11	0.12

It is thus concluded that genotypes differed significantly for various fruit characters and plant propagation studies. Out of all the 12 genotypes studied, genotype WP-3 proved to be best for various fruit and plant propagation characters followed by WP-1.

Salient Features

- The sun dried anardana from mature and ripened fruits remained fresh (in poly bags) even after 3 months of storage at room temperature, whereas in the dried seeds of immature fruits, some larvae were found which deteriorate the quality of anardana. Thus, the anardana should be prepared from mature and ripened fruits and not from unripened ones.
- Also, instead of using gunny bags or roof tops, the farmers should be encouraged to use muslin cloths for hygienic drying of anardana.
- Some small poly tent structures should also be used for aseptic drying of anardana without allowing any foreign particle to be mixed with the dried product.
- To separate seeds from the septa is a cumbersome process as it resulted in blackening or tanning of hands and thus gloves should also be used for separating seeds from rind.

Impact

Before taking up this study in 2005, *anardana* growers were selling *anardana* @ Rs. 75 per Kg but after creating awareness among the *anardana* growers in 2008, they were getting Rs. 225 per Kg by maintaining its quality. So, within 3 years their income increased three fold. Presently, they are selling it at Rs. 350-400 per kg.

Udhampur

Doda

Rajouri

Natural Plantation of Anardana.

Anardana Fruit

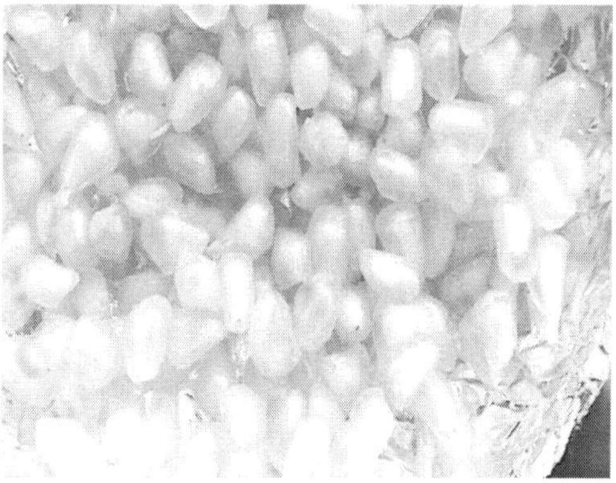

Seeds from Unripened Fruit

Seeds from Ripened Fruit

Fresh Aril

Dried Aril

Regional Agricultural Research Station, Rajouri

64. Effect of Row Ratio of Parents and Spacing on Single Cross Hybrid Seed Production of Quality Protein Maize (*Zea mays* L.)

Contact Scientist: **Dr. Vikas Sharma**

Maize (*Zea mays* L.) is an important cereal crop of the world which provides nutrition as well as photochemical compounds for human beings and animals. It is second important food crop of Jammu after rice and a major food item for the tribal and hilly population of the state. One of the prime limiting factors in increase productivity of this crop is unavailability of quality seed of improved verities/ hybrids. Quality Protein Maize contains, in general, 55 per cent more tryptophan, 30 per cent more lysine, 38 per cent less leucine and higher niacin availability due to a higher tryptophan and lower leucine content, higher calcium and carbohydrate, and carotene utilization than that of normal maize. It also provides nutritionally rich feed and fodders to poultry feed, livestock, swine and fishes. It may be helpful in poverty alleviation programme through less dependency of farmers on private seed companies for seed, easy viability of QPM cultivars hybrid seed in the local market, high economic return to the farmers and entrepreneurship development among the farming community.

Objectives

1. To find out the suitable row ratio of male and female parents ofsingle cross maize hybrid seed production for optimum yield under Rajouri conditions,
2. To find out the effect of spacing hybrid seed production for optimum yield under Rajouri conditions,
3. To work out the economics of hybrid seed production.

Methodology

A field experiment was conducted at research farm of Regional Agricultural Research Station, Rajouri of SKUAST-Jammu during *kharif* 2015 and 2016 to evaluate the suitable row ratio of male and female parents, effect of spacing for optimum yield and economics of single cross maize hybrid seed production under Rajouri conditions. The experiment was laid out in factorial randomized block design (FRBD) with three replications in 6 m x 6 m plots taking two spacing in main plots (S_1 - 60 cm x 20 cm and S_2 – 75 cm x 20 cm) and four row ratios of male and female parents (R_1 – 1:2, R_2 – 1:3, R_3 – 1:4 and R_4 – 1:5) in sub- plots of Vivek QPM- 9 *i.e.*, VQL -2 and VQL - 1. The soil of the experimental field was clay loam in texture with 190.4 kg/ha N, 8.4 kg/ha P_2O_5 and 123.2 kg/ha available K O and having a pH of 7.71. The crop was grown with the uniform fertilizer doses *i.e.*, Nitrogen -90 kg, P_2O_5 – 60 kg, K_2O – 30 kg and $ZnSO_4$ – 25 kg per hectare as per the recommendation of the university. Insect pest and disease management was done uniformly in all plots. The crop was shown on 22nd of May, 2015 and 28th of May in the year 2016, about 20 days prior to the general sowing time in this area to provide time isolation for maintaining genetic purity. The tassel of the female lines was removed as they came out from the plant to fertilize all the seed on female plants from the male parent.

Results

The results of two years pooled data revealed that row spacing S_2 (75cm x 20 cm) gave significantly higher hybrid seed yield over the row spacing S_1 (60cm x 20 cm).The same results was obtained in case of number of cobs per plant, cob length, no of grains per cob and 100 grain weight. The higher B: C ratio (1:2.25) was observed in case of row spacing S_2 over S_1 (1:1.91). The maximum hybrid seed yield (16.40 q/ha) was recorded in the row ratio of male and female parents R_2 (1:3), which was statistically superior to other treatments followed by row ratio R_3 (1:4). The similar trend was observed with respect to yield attributing characters number of cobs per plant, cob length, no of grains per cob and 100 grain weight. In case of benefit cast ratio R_2 showed maximum B: C ratio (1:2.78).There was no any interaction effect in row ratio and row spacing as per the pooled data for both the years.

Table 1: Effect of Treatments on Yield Attributing Character and Hybrid Seed Yield (F1) of QPM

Treatments	*No. of Cobs/ Plant*	*Cob Length (cm)*	*No. of Grains Cob^{-1}*	*100 Grain Wt. (g)*	*Hybrid Seed Yield (q ha^{-1})*	*B:C Ratio*
Spacing (cm)						
S_1	1.33	16.35	369.31	189.29	13.23	1.91
S_2	1.39	17.61	387.20	196.35	14.74	2.25
CD (P=0.05)	**0.12**	**0.60**	**14.63**	**NS**	1.10	——-
Row Ratio (Male : Female)						
R_1	1.38	16.68	413.18	188.63	13.48	1.97
R_2	1.37	16.67	411.88	187.46	16.40	2.78
R_3	1.28	16.34	372.77	184.10	13.76	2.18
R_4	1.16	16.45	335.85	182.58	12.20	1.69
CD (P=0.05)	**0.1**	**N.S**	**29.61**	**N.S**	1.41	——

Recommendation

Profitability of single cross hybrid seed production over grain crop showed that the enhancement in the farmer's income was about 2.78 times than the grain crop with the net return of Rs. 34,010.00 from grain yield and Rs. 94,553.00 in case of seed yield of single cross hybrid QPM. The B:C ratio of grain yield was 1.72 whereas; it was observed 2.78 in case of seed yield. The highest hybrid seed yield of single cross hybrid maize was recorded 16.40 q/ha with 1:3 ratio of male and female parents than the other ratios *i.e.*, 1:2 and 1:4 ratio.

Impact

It increases the livelihood and employment opportunities among rural people. Since the maize is a staple food of the hilly and tribal farmers so introduction of the Quality protein maize enhanced the nutritional status, an important aspect of livelihood. The high return from the seed production of single cross hybrid QPM also enhanced the economic conditions/employment opportunities of the hilly tribes of Rajouri.

Crop Stand

Detasseling or Removal of Male Flower from Female Lines

Harvested Crop of Vivek QPM - 9

Stall Visit of Hon'ble Member of Parliament, Jammu

Stall Visit of Hon'ble Member of Legislative Assembly, Nowshera, Rajouri

65. Green Fodder Production Round the Year in Rajouri District

Contact Scientist: **Dr. Vikas Sharma**

Livestock is the important subsidiary enterprise in supplementation to the crop husbandry in India. The Rajouri district has a vast and diverse livestock population. The share of livestock sector to the gross value output of the district has been increasing continuously at faster rate than the crop sector. Among these, fodder feed being the main factor in livestock production and needs to be addressed, because the feed alone constitutes 60 to 70 per cent of the production cost. Thus, any attempt towards enhancing livestock productivity should consider the feed availability. A balanced diet will keep an animal healthy and productive, it doesn't mean only purchased mineral mixtures. Instead of buying expensive concentrates, there are a variety of leguminous plants rich in protein which can be grown in the farm. Among different resources, crop residues are major one and generally defined as feedstuffs, which are bulky and contain higher fiber content (18 per cent). These are the major feed resource for livestock.

Many technological interventions we can made in multidisciplinary approach to improve the livestock productivity through increased fodder availability and accessibility.

- Cultivation from arable land
- Cultivation from non-arable land
- Management of grassland or trees

In Rajouri district there are two distinct crop seasons Kharif and Rabi. During the peak months of fodder growth is no dearth of fodder. In between the two major fodder supplying periods, there are two distinct lean periods for fodder supply

- May to mid-July
- October to December

Therefore, green fodder supply throughout the year is actually planning green fodder supply during two scarcity period in a year, which can be arranged by planning crop rotations to grow fodder for these periods.

Kharif Season

Maize

- Ideal fodder crop due to quick growing, high yielding, palatable and nutritious
- Two crops in a year under Rajouri district
- Sowing can be done March-mid September
- Ideal time of harvesting is silk to milky stage
- Maize crop yields 400 q/ha green fodder.
- Improved varieties: African tall, J-1006
- Fodder availability: April-December

Sorghum

- Main kharif fodder but is relatively slow growing and susceptible to insect and pest and diseases.
- It remains green and palatable over a longer duration than maize or bajra.
- Risk of prussic acid poisoning by feeding young plants to livestock.
- Should be harvested at flowering stage when prussic acid is quite low in concentration.
- In drought conditions, crop should be irrigated at least one week before harvesting
- Sowing time: May-July
- Improved Varieties: MP Chari, SL-44
- Green fodder yield: 600 q/ha.
- Fodder availability: July- October

Cowpea

- Quickest growing leguminous kharif fodder crop
- Sowing time: Mid March-July
- Provide balanced ration in mixture with maize, bajra
- Can be sown monoculture of mixed crop
- Gets ready for feeding in 50-55 days after sowing
- Green fodder yield: 300 q/ha.
- Improved Varieties: BundelLobia, Cowpea 88.
- Fodder availability: May-October

Rabi Season

Berseem

- Important leguminous rabi crop
- Highly nutritious, palatable fodder
- Forage availability starts just after 30 days of sowing
- Forage availability November to April
- Nearly 400 q/ha forage in 5-6 cuttings
- Can be grown as mono-crop or in mixture with oat, mustard, ryegrass.
- Improved varieties Mescavi (diploid) and Pusa Giant (tetraploid), BL-1, BL-10 Vardan and BL-22

Oat

- One of the most nutritious non-leguminous rabi crop
- Tall growing plant give one or two cuttings
- Sowing early October in higher altitudes to mid Novembers in plains
- 1st cut after 50 to 70 days of sowing and second at 50 per cent flowering
- provides 400 -500 q/ha green fodder and 10 to 14 q/ha seed if left for seed production after first cut
- Improved varieties: Kent, Sabzar, UPA 212 and Palampur 1
- Fodder availability: December-April

Lucerne

- Provides highest amount of protein per unit area
- Also provides good amount of Ca and P.
- Provides fodder round the year with slightly decrease in rainy season.
- Perennial crop, can be grown 5-7 years
- Sowing time first fortnight of October
- First cutting after 75 days of sowing
- Lucerne crop provides 500 q/ha green fodder in 7-8 cuttings.
- Improved Varieties T9 and LLC 5
- Fodder availability: Round the year but low in rainy season

Perennial Grasses

Napier Bajra Hybrid

- It is a hybrid between Napier and Bajra plant
- Vigorous, hardy, high yielding perennial hybrid fodder crop
- More jucy, succulent and less fibrous at or before harvesting

- Sowing time: Feb-April through root cuttings
- First cutting after 55-60 days.
- Green fodder: 2000 q/ha
- Improved Variety: PBN 83, CO-1, CO-3, APBN-1

Guinea Grass

- High quality multi cut kharif fodder
- Can be cultivated as annual or perennial
- Sowing time: March to May
- First cutting after 60 days and subsequent at 30 days interval
- Early growth is very slow as compared to bajra and maize
- Needs more fertilizers for giving full potential yield.
- Green fodder: 1600 q/ha
- Improved varieties: PGG-101, PGG-518
- Fodder availability: May – November

Setaria

- High quality multi cut rabi fodder
- Sowing time: March to May
- Can be cultivated as annual or perennial
- Sowing time: March to May
- Green fodder: 1800 q/ha
- Improved varieties: Nandi, PSS 1, Setaria 92
- Fodder availability: May – November

Tall Fescue

- High quality multi cut rabi fodder
- Can be cultivated as annual or perennial
- Also cultivated in shade condition
- Sowing time: March to May
- Improved varieties: HIM-1, HIMA-4
- Green fodder: 1700 q/ha

66. Rainwater Harvesting and its Demonstration for Supplemental Benefits for High Value Crops in Hill Region

Contact Scientist: Dr. Vikas Sharma

The water resources in the Rajouri district of Jammu and Kashmir state is distributed unevenly both spatially and temporary. This region is blessed to have some precipitation even during the non monsoon months; however the timing of the rains does not always coincide with the time of actual/critical requirement. Moreover, the traditionally irrigated lands in this region, which get *Kul* waters, already have the technology for distribution of irrigation waters to various fields through irrigation channels. However, the rainwater harvested in ponds is limited in quantity; so cannot be used indiscriminately like *kul* waters that are used for irrigating rice fields. In spite of sufficient rainfall in the region, availability of irrigation water during post monsoon period is always very meager. Farmers in this region have the capability of distributing the irrigation water in their fields if the water is flowing under gravity. This capability can be enhanced with the availability of harvested rainwater if they can also get access to the application technology for the harvested rainwater in ponds (*e.g.* low cost technologies like solar powered pumps for lifting the harvested rainwater).

Water harvesting structures are environmentally sound and apart from increasing bio-mass production they are helping in giving the additional benefits of flood and drought moderation, ground water augmentation, employment generation and improvement in socio-economic conditions of the people. This has resulted in collection of inevitable runoff, efficient storage of harvested water, its application and optimum utilisation for maximising production besides, supplementing the soil moisture deficiency. This region comprises of sloppy terrain and with highly percolative and impounding less water. Therefore the farmers who harvest rainwater in ponds are recommended to use it for high value crops like vegetables, flowers or fruits. Demonstrations were conducted at the farm of Regional Agricultural Research Station, Rajouri and farmers field. The farmers in Rajouri have been demonstrated the use of harvested rainwater in ponds for providing supplemental irrigation for floricultural (Village -Ladote), vegetable (Village – Doongi and Nagrota) and fruit crops (Village- Dhangri). Dug out ponds were constructed on the selected farmers' fields in these villages and the respective farmers were encouraged to use the harvested rainwater to irrigate their crops. Training activities in these villages have provided the access of the rainwater harvesting technology to many other farmers of the selected villages, who are also showing inclination to get a rainwater harvesting pond constructed in their fields.

Methodology

Field demonstrations were conducted at Regional Agricultural Research Station of SKUAST, Jammu having elevation of 920 m above mean sea level, latitude 33°37' N and 74°31'E longitude for use of harvested rainwater in ponds for providing supplemental irrigation for high value crops. The soil of the experimental area was clay loam with bulk density of 1.38g/cc. The surface soil was slightly acidic having pH 6.7 and EC of 0.084 dSm^{-1}. The farm pond having 10.5 m x 7.50 m dimensions at the bottom and 15 m x 11 m at the top with 4 m gross height was excavated. The side slopes of the pond were 1.67 H: 1 V along the length, while 1.5 H: 1 V along the width. The HDPE lining material was used as lining material. About 5 to 10 cm thick pre-lining cushioning bed of paddy straw bundles was put along all the sides and bottom so as to become it as paddy straw cushioning bed. The extended portion of the lined HDPE 250 micron film on the land surface all along the length and width of the pond buried in a trench and covered with soil on all the four sides so as to avoid entry of turbulent runoff water in to the pond. The drip was used for growing short duration vegetable crops like strawberry, broccoli, capsicum, knolkhol, cabbage, marigold and chilli. The life saving irrigation to the crop was

provided through PVC pipes preferably of about 5 cm dia uring the pre monsoon period (15 April to 15 June) and post monsoon 15th October to 15th January as and when required. If the water in the pond is not utilized for any activity, then 27 per cent of evaporation loss from the pond was estimated. For calculating the water lost from pond, the daily average of 25 years pan evaporation was taken and it was multiplied by 0.7 as to calculate the evaporation from large water bodies. The dead storage of 10 per cent of water in the farm pond was considered during the calculations

Recommendation

Demonstrations were conducted at the farm of Regional Agricultural Research Station, Rajouri and farmers field. The study reveals that the rainwater harvested in the HDPE lined farm ponds, costs 0.90 paise per litre, while the life of HDPE lining material has considered as five years. The harvested water was used for supplemental irrigation of short duration vegetable crops like strawberry, marigold,broccoli, knolkhol, cabbage and chilli and fruit crops *viz.,* nectarine and plum. The economics was also worked out and average B: C ratio was 5.40 in response to irrigation. The rain water harvesting through lined pond is viable technique for sustainable development of hilly region for livelihood security of farmers.

Benefits/Impact

After interventions and adoption of rainwater harvesting techniques helps the farmer to raise his socio-economic status and source of inspiration for other farmers of the area.

SUCCESS STORY: HTM-MM-I/2.48 (Rainwater Harvesting and its Supplemental Benefits)

Name of the farmer	Balbir Raj S/o. Sh. Diwan Chand
Address	Village Mehari Maryala, Block: Doongi District: Rajouri
Land Holding	1.5ha
Age	46 years
Education	Middle

Intervention

- The farmer was trained during on farm (SKUAST-J, RARS, Rajouri) training programmes on rain water harvesting and its use for raising high value/ cash crops.
- Motivated and guided the farmer to harvest rain water and their judicious use to cultivate high value crops *viz.,* flowers, vegetables and nursery raising of fruit, vegetable and flowers *etc.*

Output

The farmer is now carrying out integration of cereal, pulses, and oilseed crops, horticultural crops like Marigold, Strawberry, Broccoli, Cucumber, Bitter guard, Knol Khol and nurseries of different vegetable crops *etc.* in combination with fruit trees as horti-agriculture system on sustainable basis.

Outcome and Impact

Total revenue generated by the farmer

- Income from sale of flowers (Marigold)- Rs.0.75lakh/annum
- Mushroom cultivation- 0.50 lakh/annum
- Sale of nursery saplings (Flowers and vegetables) - Rs. 1.00 lakh/annum
- Income from cereals, pulses and oilseeds - Rs. 0.25 lakh/annum
- Income from sale of vegetables- Rs.0.60 lakh/annum

B. School of Biotechnology

67. Grain Quality Analysis and Molecular Characterization (DNA Fingerprinting) of Basmati Rice

Contact Scientist: **Dr. Romesh K. Salgotra**

Basmati rice is a unique extra long slender grain variety of rice grown in India and Pakistan. They are famous for their light nutty flavor, delicate texture, delicious taste, delightful fragrance and softness. They remain separate and non-sticky after cooking. They have excellent aroma, sweet taste and post cooking elongation of more than twice its original length. The grain of Basmati rice has more length than width and they grow even longer as they cook. India is the largest cultivator and exporter of this rice and the majority of Basmati rice come from the fields of Punjab. The major export destinations are Iran, Saudi Arabia, Iraq, Kewait and United Arab Emirates. The areas of Basmati rice production in India are in the states of Jammu and Kashmir, Himachal Pradesh, Haryana, Uttrakhand, Punjab, Uttar Pradesh and Delhi. The quality of basmati rice is based on a variety of physical and chemical properties such as head rice and broken grains, long, medium and small grains. Foreign material like dirt, stones, shape and zize of grains, damage and discolored grains, colour, texture, aroma, moisture content, chalkiness *etc.* In agriculture industry, quality of grains is an important issue to protect the consumers from substandard products because the samples of the material are subjected to adulteration.

Objectives

- Determination of basmati rice grain and seed quality characteristics
- Molecular characterization (DNA fingerprinting using SSR markers) of basmati rice.

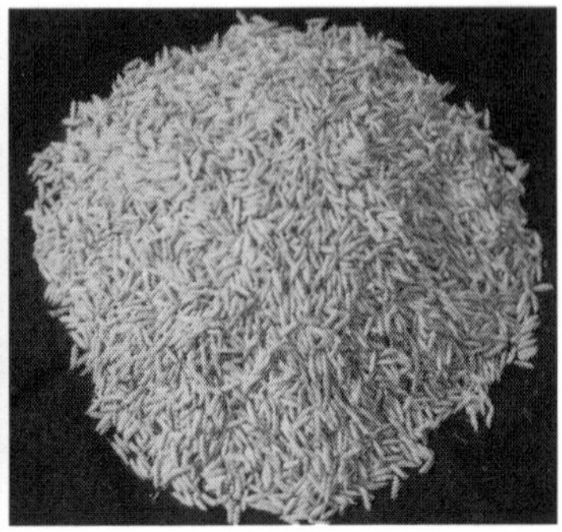

Grain Quality Characteristics

Determination of following basmati grain quality:

- Shape
- Size
- Color of grain
- Chalkiness
- Admixture
- Grain length (L)
- Grain breadth (B)
- L/B ratio
- KLAC
- Gel consistency
- Moisture
- Alkali spreading value (ASV)
- Aroma

Molecular Characterization (DNA fingerprinting) of Basmati Rice

Molecular markers allow us to make estimates of genetic diversity directly at the DNA/molecular level, reducing the interference of environmental variations. Molecular markers can also serve as a tool for variety/genotype identification. Molecular markers involve any kind of marker system that differentiates two individuals at the molecular level.

Methodology

A. DNA Extraction

For DNA extraction fresh and young leaf is the preferred tissue due to low concentration of metabolites and polysaccharides. There are many published DNA extraction techniques having common features. Major steps include:

Protocol

1. Collect healthy leaf sample, wash in sterile water and dry on paper towels.
2. Cut the leaf tissue into small segments and place the tissue in a sterile mortar.
3. Add liquid nitrogen inside mortar and crush with sterile pestle until completely ground.
4. Once leaf tissue is in powdered form, transfer the powder to microcentrifuge tube with the help of sterile spatula and immediately add 750 µl of pre-warm DNA extraction buffer using micropipette.
5. Mix and transfer the entire content in to a sterile microcentrifuge tube.
6. Incubate at 60°C for 1 hr in shaking waterbath.
5. Remove the tubes from water bath and add 750 µl of CI (choloroform: isoamyl alcohol, 24:1)
6. Mix the contents and spin for about 10 min at 12,000 rpm in a centrifuge at 4°C. Transfer the top (aqueous) phase (~500-600 µl) to another labeled microcentrifuge tube using a 200 µl capacity micropipette using cut tips (tips cut at the bottom using sterile scissors) without disturbing the interphase.
7. To this add equal volume of chilled isopropanol, mix well and store in -20°C for 0.5hr.
8. Spin for 10 min at 12,000 rpm. DNA will settle at the bottom as a pellet after centrifugation. Decant and discard the supernatant.
9. To the pellet, add 200 µl of 70 per cent ethanol (70 per cent EtOH and 30 per cent sterile double distilled water, V/V), gently tap the pellet to release it from the tube, centrifuge at 12,000 rpm for 5 min and decant the supernatant.
10. Keep the microcentrifuge tubes containing the DNA pellet open at room temperature (air drying) for about 1-2 hours.
11. After the pellet is completely dried (with no smell of alcohol coming from the tube), resuspend the DNA pellet in 100-200 µl of TE buffer (10 mM Tris-HCl, pH 8.0 and 1 mM EDTA, pH 8.0 prepared using sterile double distilled water). Store the DNA at –20°C.

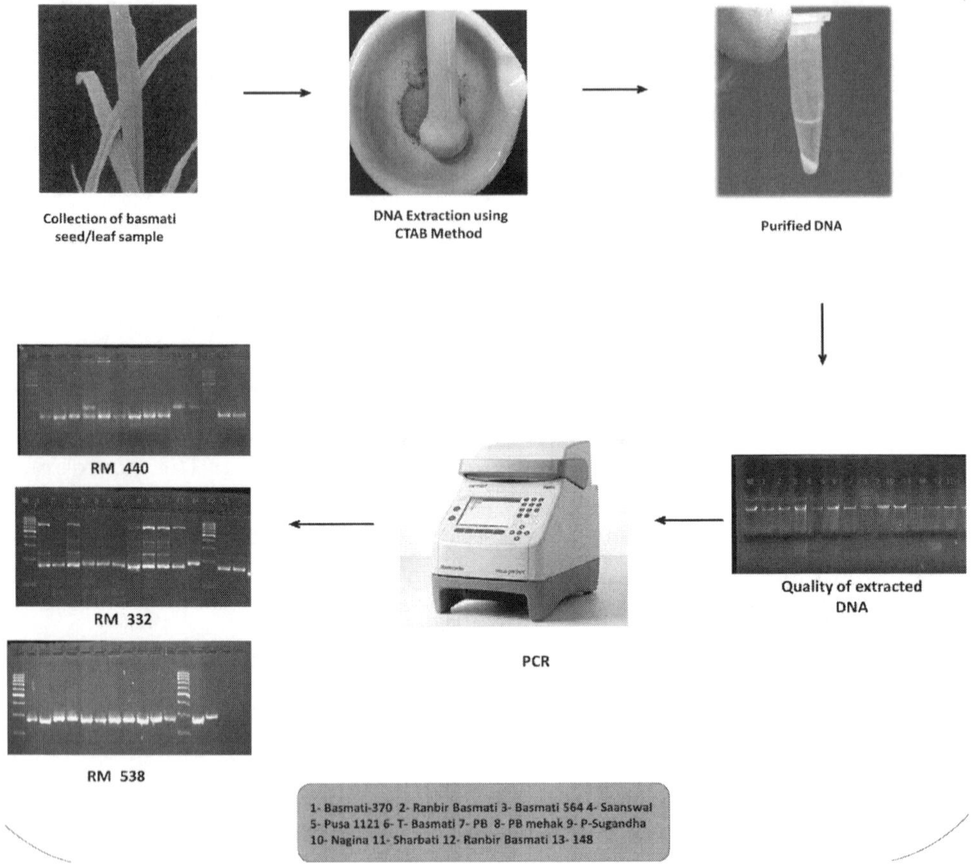

Figure 1: DNA Fingerprinting of Basmati Rice Varieties using SSR Markers.

DNA Quantification

DNA can be quantified by various methods. These include:

1. Agarose gel electrophoresis for comparison of extracted sample with standard DNAs of known concentration
2. Spectrophotometric determination; now a days Nanodrop™ or other automated systems are used in which 1-2 µl of DNA can be quantified as per manufacturer's instructions. It is automated and gives A_{260}/A_{280} ratios (which determines the quality of DNA) and concentration of DNA and accordingly dilutions can be made.

B. Microsatellite Markers/Simple Sequence Repeats (SSR)

The analysis of simple sequence repeats (SSRs or microsatellites) using the polymerase chain reaction (PCR) is now widely used and has several advantages. SSR markers are expressed codominantly, are generally robust, repeatable, easy to score (record), and easy to automate for high-throughput detection. Moreover several SSRs can be multiplexed and detected by nonradioactive DNA sequences or on gel electrophoresis or capillary electrophoresis. With recent advancements in genomics mapped SSRs are now available in several crops. For effective microsatellite analysis, selecting high quality SSR markers that are unlinked, polymorphic, reliable and reproducible and having no stuttering is essential. Primers for any kind of marker system should be synthesized by an assured supplier. Some SSRs are publicly available.

C. Polymerase Chain Reaction (PCR)

The polymerase chain reaction is cárried out in a laboratory apparatus called as thermal cycler or PCR machine. The device has a thermal block with holes where tubes holding the above reaction mixtures are inserted and the cycler raises and lowers the block temperature in discrete, pre-programmed temperature cycling steps.

Various steps in the polymerase reaction include: denaturation, annealing and extension.

i. **Denaturation** is the step in which 2 strands of DNA get separated from each other. It is usually carried out at 94°C to completely denature DNA.
ii. **Annealing** is the step in which primers get attached to single stranded DNA. The optimal annealing temperature is 5°C lower than the melting temperature (T_m) of primer-template DNA duplex. T_m can be roughly calculated as 4(G+C) + 2 (A+T).
iii. **Extension** is generally performed at 72°C. In this step synthesis of complementary DNA strand takes place in which dNTPs are added with the help of DNA polymerase (*Taq* polymerase)

These 3 steps are repeated 30-40 times to get good amplification.

Protocol for PCR Amplification

1. Take out $MgCl_2$, dNTPs, primers and buffer from the deep freezer and let these thaw at room temperature. *Taq* DNA polymerase needs to be taken out only when it is to be added and return immediately it into the deep freezer after use.
2. For setting up PCR reaction, prepare master mix in 2.0 μL centrifuge tube of the following components except template DNA. Keep all these components on ice in ice bucket.

Sl.No.	Component	Stock Concentration	Required Concentration	Amount Required for One Reaction
1	Taq polymerase buffer	-	1X	2.5μL
2	$MgCl_2$	25mM	3mM	3.0μL
3	Primer	5μM	0.5μM	2.5μL
4	dNTPs	25mM	0.2mM	0.2μL
5	Taq DNA polymerase	5U/μL	1U	0.2μL
6	Template DNA	10ng/μL	50ng	5.0μL
7	Sterile double-distilled water (to make 25 μL volume)	-	-	11.6μL
	Total volume			25.0 μL

D. Agarose Gel Electrophoresis

It is performed in similar manner as described under section ——B (DNA quantification by gel electrophoresis). However, the agarose concentration may vary from marker to marker. The amplified fragments are separated electrophoretically and banding patterns are detected by different methods such as staining and autoradiography.

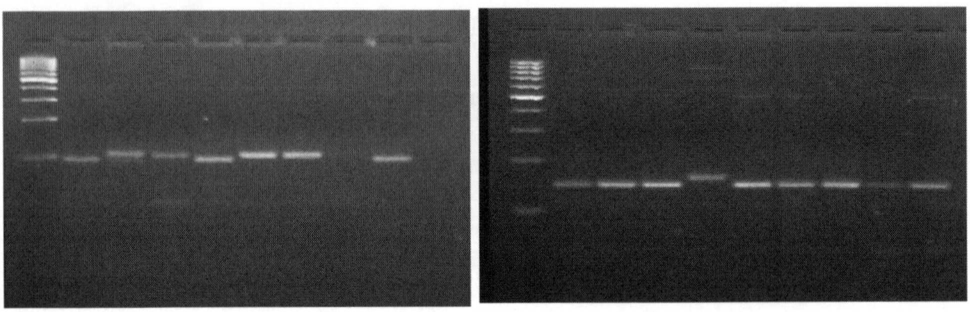

Figure 2: Bands Observed after Agarose Gel Electrophoresis.

E. Scoring and Data Analysis

The molecular data obtained is in the forms of bands of different molecular weight sizes. Based on the size and migration of bands on a gel, the data is recorded in the form of molecular weight sizes. The presence '1' or absence '0' of a band of a particular molecular weight is scored as two alleles at a single locus. This data is used to construct a binary matrix on an excel sheet. The data is used to evaluate similarity, gives a rough estimate for variation and the distances between genotypes. Data is fed to various software packages with different analytical methods that can either be downloaded from internet or purchased. These include NT-SYS, STATISTICA, DARWIN, Phylip, POPGENE *etc.*

Benefits

Basmati is an aromatic long-grain type of rice (*Oryza sativa* L.) cultivated exclusively in the foothills of the Himalayas stretching across the Jammu and Kashmir, Punjab, Haryana and Uttar Pradesh regions of India and the Punjab province of India and Pakistan. Among the basmati varieties traded in the international market, traditional basmati varieties such as Basmati 370, Dehradun, and Taraori command higher price compared to evolved basmati varieties such as Pusa Basmati, Mahi Sugandha. It is difficult to differentiate between traditional basmati, evolved basmati, and non basmati rice grains on the basis of visual observation or physicochemical tests, unscrupulous traders adulterate traditional basmati samples with evolved or non basmati grains to earn profits. Hence, distinguishing traditional basmati rice varieties from evolved basmati and look-alike long-grain non-basmati varieties has become an important aspect of the basmati trade. The suitability of microsatellite markers for basmati genotyping and adulteration detection has been demonstrated earlier. These studies have enabled Basmati importers and exporters to move toward a uniform and strict regime of DNA-based testing of the export lots.

68. Molecular Characterization: A Tool for Diversity Analysis and Varietal Identification in Vegetable Crops using DNA Markers

Contact Scientist: Dr. Susheel Sharma

Molecular markers allow us to make estimates of genetic diversity directly at the DNA/molecular level, reducing the interference of environmental variations. Molecular markers can also serve as a tool for variety/genotype identification. Molecular markers involve any kind of marker system that differentiates two individuals at the molecular level. DNA markers popularly termed as molecular markers belong to 4 major categories based on:

1. Southern hybridization, *e.g.* Restriction Fragment length Polymorphism (RFLP).
2. Polymerase Chain Reaction, *e.g.* Random Amplified Polymorphic DNA, Simple Sequence Repeats (SSRs) and Inter Simple Sequence Repeats (ISSRs) *etc.*
3. Hybridization and PCR *e.g.* Amplified Fragment Length Polmorphism (AFLP), Cleavage Amplified Polymorphisms (CAPs).
4. DNA sequence information *e.g.* SNP

Methodology

For carrying out molecular characterization of various vegetable genotypes following major steps are employed:

A. DNA extraction
B. DNA quantification
C. PCR amplification with a suitable marker system
D. Agarose gel electrophoresis
E. Scoring of data and molecular analysis

A. DNA Extraction

For DNA extraction fresh and young leaf is the preferred tissue due to low concentration of metabolites and polysaccharides. There are many published DNA extraction techniques having common features. Major steps include:

a) Tissue lysis (disruption of the tissue)
b) Subjecting DNA in to suitable extraction buffer and deproteinization
c) DNA precipitation
d) DNA purification

Materials and Equipments Required

1. Pestle and mortar (autoclaved and dried)
2. Micropipettes (1 ml, 20 and 200 ml range)
3. Spatula
4. 2 ml microcentrifuge tubes (autoclaved and dried)
5. 1 ml and 200 ml tips (autoclaved and dried)
6. Shaking waterbath
7. Refrigerated Centrifuge
8. -20°C deep freezer

Protocol

1. Collect healthy leaf sample, wash in sterile water and dry on paper towels.
2. Cut the leaf tissue into small segments and place the tissue in a sterile mortar.
3. Add liquid nitrogen inside mortar and crush with sterile pestle until completely ground.
4. Once leaf tissue is in powdered form, transfer the powder to microcentrifuge tube with the help of sterile spatula and immediately add 750 ml of pre-warm DNA extraction buffer using micropipette.
5. Mix and transfer the entire content in to a sterile microcentrifuge tube.
6. Incubate at 60°C for 1 hr in shaking waterbath.
5. Remove the tubes from water bath and add 750 ml of CI (choloroform: isoamyl alcohol, 24:1)
6. Mix the contents and spin for about 10 min at 12,000 rpm in a centrifuge at 4°C. Transfer the top (aqueous) phase (~500-600 ml) to another labeled microcentrifuge tube using a 200 ml capacity micropipette using cut tips (tips cut at the bottom using sterile scissors) without disturbing the interphase.
7. To this add equal volume of chilled isopropanol, mix well and store in -20°C for 0.5hr.
8. Spin for 10 min at 12,000 rpm. DNA will settle at the bottom as a pellet after centrifugation. Decant and discard the supernatant.
9. To the pellet, add 200 ml of 70 per cent ethanol (70 per cent EtOH and 30 per cent sterile double distilled water, V/V), gently tap the pellet to release it from the tube, centrifuge at 12,000 rpm for 5 min and decant the supernatant.
10. Keep the microcentrifuge tubes containing the DNA pellet open at room temperature (air drying) for about 1-2 hours.

11. After the pellet is completely dried (with no smell of alcohol coming from the tube), resuspend the DNA pellet in 100-200 ml of TE buffer (10 mM Tris-HCl, pH 8.0 and 1 mM EDTA, pH 8.0 prepared using sterile double distilled water). Store the DNA at –20°C.

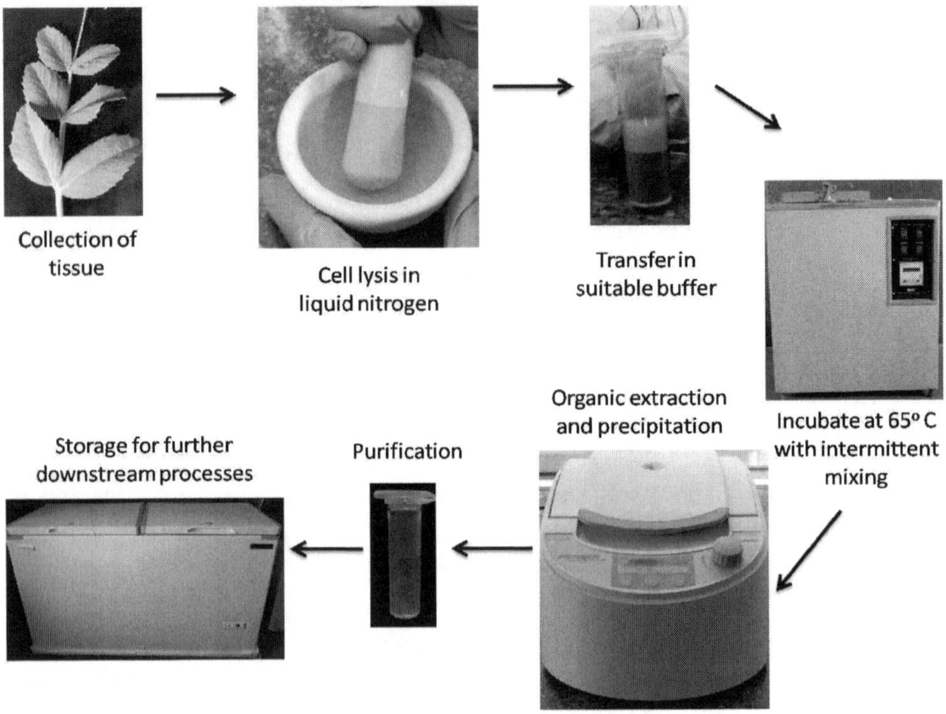

Figure 1: Different Steps of DNA Extraction.

B. DNA Quantification

DNA can be quantified by various methods. These include:

i) Agarose gel electrophoresis for comparison of extracted sample with standard DNAs of known concentration

ii) Spectrophotometric determination; now a days Nanodrop™ or other automated systems are used in which 1-2 ml of DNA can be quantified as per manufacturer's instructions. It is automated and gives A_{260}/A_{280} ratios (which determines the quality of DNA) and concentration of DNA and accordingly dilutions can be made.

Agarose Gel Electrophoresis

1. To prepare 100mL of a 0.8 per cent agarose solution, measure 0.8g agarose and add 100ml 1X TAE (Tris-Acetic acid-EDTA).
2. Microwave until agarose is dissolved and solution is clear.

3. Allow solution to cool to about 55°C and add ethidium bromide to a concentration of 0.5 μg/ml.
4. Place the comb in a gel tray.
5. Pour gel solution into tray and allow the gel to solidify for 30 min at room temperature.
6. To run, gently remove the comb, place the tray in electrophoresis chamber, and cover (wells should be submerged) with TAE buffer.
7. To each DNA sample (take 2-3 μL on parafilm), add 3 μL of 1X gel loading dye solution. Mix well both. Load 5-6 μl of sample per well. Also load DNA ladder for size determination.
8. Connect the electrodes to the power pack and carry out electrophoresis at 50-150 Volts until the sample has reached three fourth of the gel length.
9. Examine the gel under UV light in the gel documentation system and obtain a photograph.
10. Compare the intensity of the DNA bands of the samples with the intensity of the standard DNA bands. As the amount of DNA present in each standard DNA bands is known, the amount of DNA of each sample can be calculated by comparing the fluorescent yield of the sample with that of the standard.

Note: Ethidium bromide is a mutagen and a probable carcinogen. Wear gloves when handling ethidium bromide solutions. UV light can cause burns and can damage the eyes.

Storage of DNA

DNA can be stored in concentrated form at 4°C for quite a long time (few to 12 months) without affecting the quality. For long term storage, it is recommended to store at –20°C. DNA in diluted form should be strictly stored at 4°C only, as frequent thawing and freezing leads to physical degradation of DNA and drastic reduction in quality.

C. PCR Amplification with a Suitable Marker System

Polymerase Chain Reaction (PCR)

The polymerase chain reaction (PCR) developed by Kary B. Mullis in mid 1980's is a rapid procedure for *in vitro* enzymatic amplification of a specific segment of DNA, the ingredients of which include:

a) DNA template (double stranded DNA which is to be amplified);
b) primers (two oligonucleotide sequences flanking the region of interest);
c) DNA polymerase (eg. *Taq* polymerase);
d) a mixture of deoxyribonucleoside triphosphates (dNTPs); and
e) an appropriate buffer and salts

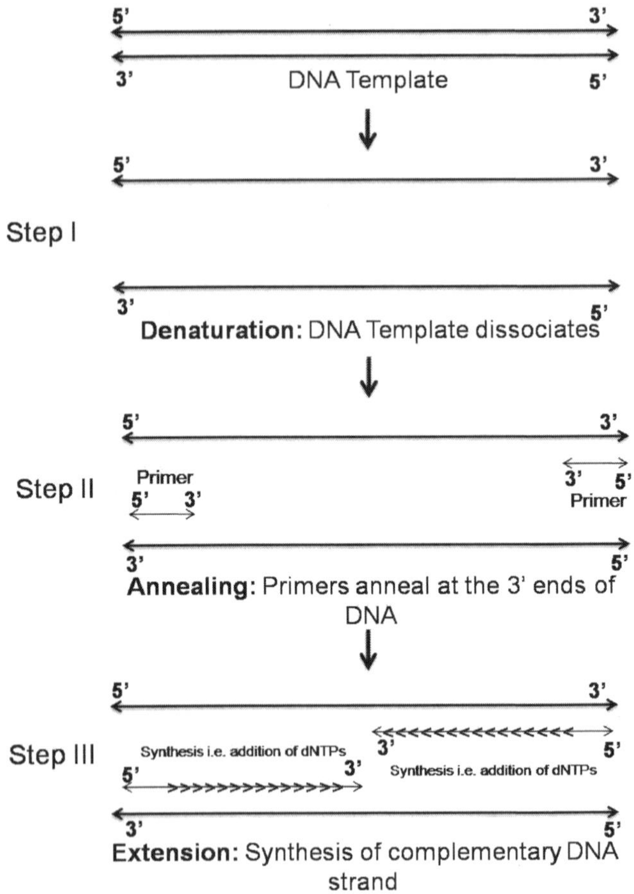

Figure 2: Steps of Polymerase Chain Reaction (PCR).

The polymerase chain reaction is carried out in a laboratory apparatus called as thermal cycler or PCR machine. The device has a thermal block with holes where tubes holding the above reaction mixtures are inserted and the cycler raises and lowers the block temperature in discrete, pre-programmed temperature cycling steps.

Various steps in the polymerase reaction include: denaturation, annealing and extension.

i. **Denaturation** is the step in which 2 strands of DNA get separated from each other. It is usually carried out at 94°C to completely denature DNA.
ii. **Annealing** is the step in which primers get attached to single stranded DNA. The optimal annealing temperature is 5°C lower than the melting temperature (T_m) of primer-template DNA duplex. T_m can be roughly calculated as 4(G+C) + 2 (A+T).

iii. **Extension** is generally performed at 72°C. In this step synthesis of complementary DNA strand takes place in which dNTPs are added with the help of DNA polymerase (*Taq* polymerase)

These 3 steps are repeated 30-40 times to get good amplification.

Choice of Molecular Markers

A number of PCR based markers are presently being used for diversity studies and the most commonly used include RAPDs, ISSRs and SSRs. The choice of marker used depends on the availability of sequence information. If genome sequence of vegetable crop is available then SSRs are used. On the contrary, lack of genome sequence information necessitates using RAPDs and ISSRs. Now-a-days, due to next generation sequencing (NGS) techniques, sequencing of many vegetable crops including tomato, cucumber, cassava *etc.* have been carried out and are available in public domain (Van *et al.*, , 2013). The sequence information can be used for designing and use of SSR markers for diversity studies and other crop improvement purposes. Whatever the marker system is used, the basic methodology employed remains the same. The basic steps for RAPDs, ISSRs and SSRs are (i) DNA extraction (ii) DNA amplificationusing specific/randomprimers (iii) separation of amplified fragment on agarose gel/acrylamide gel and (iv) visualization of markers on gel.

Random Amplified Polymorphic DNA (RAPD)

Random amplified polymorphic DNA (RAPDs) were the first and most common PCR-based molecular markers to be employed in genetic variation analyses (Williams *et al.*, , 1990). RAPD markers are generated through the random amplification of genomic DNA using short arbitrary and single primers (decamers) acting as forward and reverse primers. Polymorphisms are detected as the presence or absence of bands and mainly results from sequence differences in the primer binding sites. The enormous attraction of these techniques is that, there is no requirement for DNA probes or sequence information for primer design thus; it can be employed across species using universal primers.

Inter Simple Sequence Repeat (ISSR)

Inter simple sequence repeat (ISSR) technique is a PCR based method, which involves amplification of DNA segment present at an amplifiable distance in between two identical microsatellite repeat regions oriented in opposite direction. The technique uses microsatellites, usually 16–25 bp long, as primers in a single primer PCR reaction targeting multiple genomic loci to amplify mainly the inter-SSR sequences of different sizes.

Microsatellite Markers/Simple Sequence Repeats (SSR)

Microsatellites are the simple sequence tandem short oligonucleotide repeats distributed throughout the genome. PCR-based assays for such SSRs have become the most popular and powerful of the current methods for genotype identification as they are highly polymorphic, reproducible and co-dominant mendelian markers.

PCR primers flanking to the SSR can be developed and variation can be detected on the basis of PCR amplification, separating amplified products on poly acrylamide gel electrophoresis (PAGE) and bands can be obtained by silver staining of the gels. However, a high resolution agarose, Metaphor ® can also be used.

Materials and Equipment required for PCR

1. Thermal cycler
2. Micropipettes
3. PCR tubes (0.2 or 0.5 μL)
4. Spinnet
5. Centrifuge tube (2 mL)
6. Ice bucket having ice flakes
7. PCR reaction components (as given in tabular form)

Protocol for PCR Amplification

1. Take out $MgCl_2$, dNTPs, primers and buffer from the deep freezer and let these thaw at room temperature. *Taq* DNA polymerase needs to be taken out only when it is to be added and return immediately it into the deep freezer after use.
2. For setting up PCR reaction, prepare master mix in 2.0 μL centrifuge tube of the following components except template DNA. Keep all these components on ice in ice bucket.

Sl.No.	*Component*	*Stock Concentration*	*Required Concentration*	*Amount Required for One Reaction*
1	Taq polymerase buffer	-	1X	2.5μL
2	$MgCl_2$	25mM	3mM	3.0μL
3	Primer *	5μM	0.5μM	2.5μL
4	dNTPs	25mM	0.2mM	0.2μL
5	Taq DNA polymerase	5U/μL	1U	0.2μL
6	Template DNA	10ng/μL	50ng	5.0μL
7	Sterile double-distilled water (to make 25 μL volume)	-	-	11.6μL
	Total volume	25.0 μL		

* Incase of SSRs separate forward and reverse primer is used so concentration will be halved accordingly.

3. Switch on the thermocyler and place samples in it. The following general thermocycling steps are followed:

Step1: Initial denaturation at 94°C for 5.0min

Step2: Denaturation at 94° C for 1.0min

Step3: Primer annealing at 45-55° C for 1.0min (Primer annealing temperature may vary)

Step4: Primer extension at 72° C for 2.0min

Step5: Go to 2, 39 times

Step6: Final extension at 72° C for 10min

Step7: 4° C for ever.

4. After the run is completed turn off the machine and remove the samples and store in refrigerator.

D. Agarose Gel Electrophoresis

It is performed in similar manner as described under section ——B (DNA quantification by gel electrophoresis). However, the agarose concentration may vary from marker to marker. The amplified fragments are separated electrophoretically and banding patterns are detected by different methods such as staining and autoradiography.

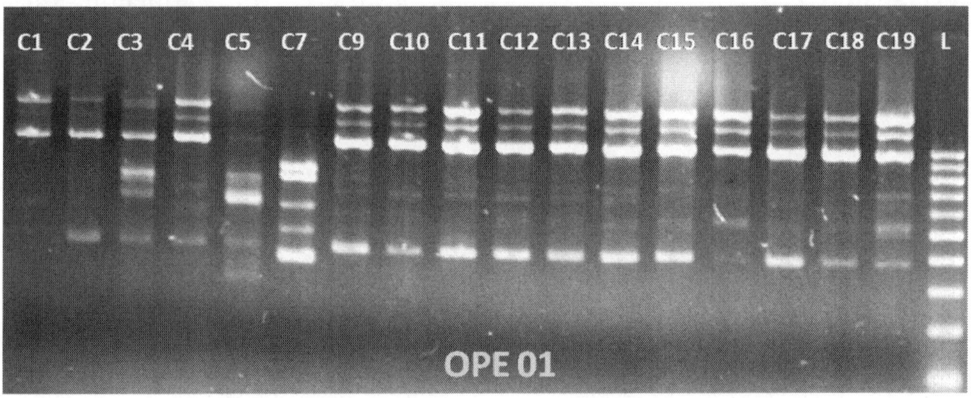

Figure 3: Typical RAPD Pattern of Primer OPE 01 as Applied in some Genotypes of Cucumber.

E. Scoring and Data Analysis

The molecular data obtained is in the forms of bands of different molecular weight sizes. Based on the size and migration of bands on a gel, the data is recorded in the form of molecular weight sizes. The presence '1' or absence '0' of a band of a particular molecular weight is scored as two alleles at a single locus. This data is used to construct a binary matrix on an excel sheet. The data is used to evaluate similarity, gives a rough estimate for variation and the distances between genotypes. Data is fed to various software packages with different analytical methods that can either be downloaded from internet or purchased. These include NT-SYS, STATISTICA, DARWIN, Phylip, POPGENE *etc.* Finally, a dendrogrm (tree) and a similarity matrix are generated. The same data can also be used for identification of genotypes which helps in either protection of breeder's rights or for quality control of F_1/cultivar production or for identification of duplicates. For this, the easiest approach is to identify genotypes with unique bands. However, some statistics can also be used in genotype identification particularly for the efficacy of either specific primers or DNA marker techniques. These functions include polymorphism information content (PIC), marker index, resolving power, discriminating power, *etc.*

Benefits

Assessment of genetic diversity on morphological basis is very difficult due to increasing number of collections of vegetable germplasm. Moreover, there are certain drawbacks associated with morphological markers like epistatic interactions, pleiotropic effects and influence of environment on trait expression. Studying genetic variation through modern technology involving molecular markers has played a tremendous role in delineating the diversity of many vegetable crops. Molecular markers have dramatically increased knowledge and ability to characterize genetic diversity in the germplasm pool for any crop species including vegetable crops. These can also serve as a tool for variety/genotype identification. The DNA marker systems have given substantial impetus to genetic diversity studies in vegetable crops and marker assisted selection (MAS) programmes.

C. Faculty of Basic Sciences, SKUAST Jammu, Chatha

Division of Plant Physiology

69. Management of Unfruitfulness in Olive

Contact Scientist: **A.K. Tiku**

Introduction

Olive (*Olea europaea*) commonly known as zaitoon, is an attractive evergreen tree with grey-green foliage. Olive is a native to Mediterranean region (Zeven and Zhukovsky, 1975), tropical and central Asia and various parts of Africa. It has a history almost as long as that of western civilization. Its mention has been made in Holy Quran and Bible. It is regarded as a symbol of prosperity and peace. Commercial production of the crop occurs within latitudes 30 to 45 in both northern and southern hemispheres and is favored by climates characterized by mild winter, and warm summers (Sanz -Cortes *et al.*, , 2002).

Methodology

Crop Production Technology

Planting

Necessary cultivation work may be carried out before planting, such as uprooting of other trees and bushes and leveling of the soil. Planting is done into holes/pits that can be dug manually in dimensions of 60x40cm or 100x100cm. The trees are planted together with the root ball and the pit is then filled with the soil.

Trees are usually planted in the field after two years in the nursery when the main stem attains 1.5 to 3.0 cm thickness. The planting of olive in the intermediate hilly zone is done in the month of December when they are least active. The olive varieties available at Advanced Centre for Horticulture Development, Govindpura, Ramban include-*Leccino, Frontoio, Corotina, Pendolino, Zaituna, Ciprisino, Cornicobra, Itrana, Belice, Etnea, Tonda iblea, Picholine, Ottobratica, Moraiolo, Carolea, Messinese, Bioncolilla, Carigonola and Azapa.*

Planting Distance

Olive tree planting lay out depends upon the system of cultivation *i.e.* intensive/ non- intensive. In rainfed areas having poor soils, a planting distance of 6 x 4 m is ideal (Verma and Singh, 2002).

Basin Preparation and Terracing

In the intermediate zone, the land available for olive orchard establishment will mostly be sloppy. It is important that the sloppy land should be terraced and basins prepared around the trees. This will prevent the run-off of water and therefore, there will be less leaching of nutrients.

Training and Pruning

The olive trees are sometimes left unpruned for 2 to 3 years. Bearing trees are lightly pruned and crowns are thinned occasionally to permit the entry of sun light. Since the intensity of pruning affects fruit yield, care is to be taken to remove sprouts, injured, diseased, interlocked and dried branches. Besides, light pruning is done every year so as to keep the trees in active vegetative growth. During 1st two years after planting, the fast growth period of the trees, require specific pruning to maintain their growth and to keep them in good health.

Basin Preparations and Terracing.

Main Points

1.Too much pruning at young age will stunt the tree growth. 2. Acentral leader trunk will assist growth with early stages. If young tree is 3 feet tall, has side branches all the way up its trunk, do not take off all of them. a) At first, only take off any that are growing below 1foot and then in several months time when the tree has grown considerably more on top and has recovered from first pruning, take off any branch between 2 feet. Repeat this process until finally after about 2 years; you have a clear straight trunk for 40-48 inches. Do not prune too much too soon, so as not to put the tree into shock. b) Always leave large number of leaves on the tree for maximum photosynthesis to have the maximum root growth *etc.* c) Choose four evenly spaced lateral branches which may not come from exactly same height but should not be lower than 30 inches from the ground. Finally we have a young tree with approximately 4 main scaffold branches. The shape is commonly known as inverted conical vase.

Pruning for Fruiting

Olive trees produce fruits in previous year branches. Very vigorous branches are not productive (they have only vegetation buds) and week branches produce few fruits. For this reason, the aim of pruning is to induce branches to form fruits, ensure good lighting conditions and maintaining the fruiting zone active and vigorous (Metzidakis, 1997). In case of table olive varieties, pruning must also improve the size of the fruits.

Alternate Bearing Phenomenon

Olive trees historically have been notorious for their tendency to alternate a heavy crop with a light crop. The physiological basis for this condition is that excessive flowering and fruiting literally exhaust the tree. A heavy crop removes much of the various carbohydrate materials, organic substances and other essential nutrients so that adequate stored food reserves are not available for production of a crop in the following year. Maturing the heavy crop may so weaken the tree so much so that it does not even bloom the following spring.

Chemical Thinning

NAA 100-150 ppm during 0.5- 2.5 weeks from fruit set.

Hand Thinning

Removing 5-6 fruits from each branch, leaving 3- 5fruits/30 cm on the branch.

Why thin your Olive Crop

- For larger fruit - overloaded plants have smaller fruit and not of caning size
- Higher quality fruit - have higher flesh to pit ratio and higher oil content
- More consistent yearly crops

- Early maturity
- Lower harvest cost

Irrigation Requirements

The irrigation should be applied at critical stages of growth. One before flowering to minimize the premature pistil abortion, second at the time of fruit set to avoid excessive fruit drop, and third should be applied at the time of maturity to reduce the preharvest fruit drop. Also two irrigations after harvest are essential.

Rainwater Harvesting Tank.

Nutrition

To maintain the trees in good health and to harvest quality fruit, it is essential to apply the recommended doses of farm yard manure (FYM) and inorganic fertilizers especially N, P and K. Nitrogen is applied in three split doses and P and K in the month of December/January along with the first split dose of N. Second dose of N is applied before flowering and third dose at pit hardening stage. Also apply borax @ 200 g per tree in bearing trees once in alternate years (Chadha, 2001). Before plantation it is important to have the nutrient analysis of the soil done. The studies on the effect of different levels of N, P and K on the growth and nutrient content of leaves revealed that the results are in conformity with the fertilizer schedule recommended by Chadha (2001).

Manuring and Fertilizers Schedule per Tree

Age (years)	*FYM (kg)*	*N (g)*	*P (g)*	*K (g)*	*Urea (g)*	*SSP (g)*	*MOP (g)*
01	10	75	50	50	160	300	80
02	15	150	100	100	320	600	160
03	20	225	150	150	480	900	240
04	25	300	200	200	640	1200	320
05	30	375	250	250	800	1500	400
06	35	450	300	300	960	1800	480
07	40	525	350	350	1120	2100	560
08	45	600	400	400	1280	2400	640
09	50	675	450	450	1440	2700	720
10 and above	50	750	500	500	1600	3000	800

Source: Chadha, 2001.

Spray of 0.5 per cent Borax Solution during Flowering.

Fruit Set and Fruit Retention

For effective fruit set and fruit retention, it is recommended to have the foliar spray of the trees with 0.5 per cent borax solution 20 days after the emergence of first inflorescence (Tiku, 2006).

Pollinators

Olive varieties may be self compatible or self incompatible. It has been observed that cross pollination increases the fruit production and pollinators produce more consistent crops. Plant or graft a pollinator row every fourth row in the orchard.

Harvesting

The harvesting of olive fruit costs >50 per cent of the gross return and this is the major factor in increasing the production costs. With smaller acreages experiment with hand held picking aids. To have the quality olive oil production it is important that the fruits are not harvested too early or too late as that would in the former case lead to loss of fruit size, weight and oil and in the latter case loss of quality, fruit drop and reduce next years flowering.

Full Blooms

Mature Fruits

Ripened fruits

IPM Strategies in Olive Orchard

1. Light winter pruning of the trees provide open, airy trees which discourage black scale infestation and is preferred to chemical treatment. In addition, biological control is effective, since a number of natural enemies, including both parasites and predators attack black scale.

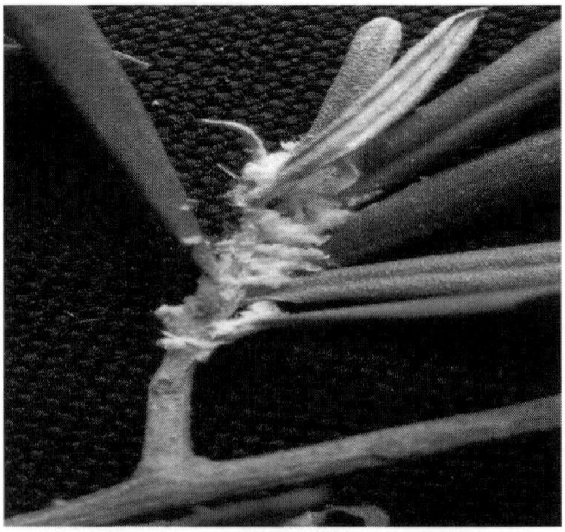

Psylla Infected Branch of Olive.

2. Tilling the area under the tree and in entire olive orchard after harvest helps in killing the pupae of olive fruit fly. Management includes bait sprays, trapping of adult flies, harvest timing, fruit sanitation after harvest, and conservation of biological control agents. Small plywood rectangles dipped in 0.1 per cent aqueous solution of deltamethrin for 15 minutes and added to bait stations containing either sex pheromone or ammonium bicarbonate, a food attractant, is effective in controlling fruit flies.
3. For sap sucking insects like tinged bug, olive psylla and black scale the following spray schedule should be followed:

Insecticides	*Concentration*	*Spray*	*Time*
Monocrophos/ imidacloprid	0.04 per cent	1st spray	First week of February
	0.005 per cent	2nd spray	Before flowering
		3rd spray	Six weeks after fruits set
		4th spray	After fruit harvest

4. For twig borer, alternate spray of monocrotophos followed by cypermethrin @1 ml per litre of water. The pest is satisfactorily controlled by removing the webbing of frass followed by plugging the holes with cotton wool soaked in the 2-5 ml of dichlorvos 76 per cent EC solutions (20 ml/10 litres of water). The holes are then sealed with mud.

Recommendations

- Studies on the on-farm trial of olive revealed that for successful flowering and fruitfulness, accurate nutritional and water management together with selective pruning are the primary tools to have a good flower population.
- For fruit set retention, 0.5 per cent borax solution is advocated for spraying twenty days after emergence of first inflorescence.

Benefits/Impact

- Non-bearing of the olive orchards has been a problem confronting the farmers and the state Government.
- To make commercially viable vast stretch of wild olive plantation (10 thousand hectares) in the intermediate hilly zone of Jammu.

D. Krishi Vigyan Kendras

---

70. Mitigation of Stripe Rust (Yellow Rust) Disease of Wheat through Chemicals and Varieties

***Contact Scientist:* Dr. Shahid Ahamad**

Stripe rust of wheat (*Puccinia striiformis)* is a major constraint for wheat production in Jammu region. There are many curative measures that can control these fungal diseases. However, there are some difficulties *viz.* cultural practices may be difficult to implement and chemical approaches may be accompanied by detrimental effects. Genetic resistance is the most efficient and economical approach to mitigate the losses caused by stripe rust diseases. However, in order to achieve genetic resistance, it is necessary that wheat lines carrying genes imparting resistance against these diseases are identified. This can either be achieved by classical phenotyping approaches such as multi-pathotype testing for stripe rust (*Sr*) genes or by gene postulation, both of which is time consuming as well as labor intensive.

In India, stripe rust was observed in Punjab, Haryana, Western Uttar Pradesh and Jammu and Kashmir and was adjudged to be responsible for considerable yield losses. During cropping season, the stripe rust appeared in severe form in almost all the wheat growing areas of Jammu region, wherein, wheat was grown over an area of 239 thousand hectares with a production and productivity of 465.33 thousand tones and 19.47q ha^{-1}. As most of wheat varieties growing in Jammu regions have become susceptible to yellow rust disease, there is a need to know source of resistance in present wheat varieties against the stripe rust.

Methodologies

Present investigations was carried out for two consecutive seasons (Rabi) during 2014-15 and 2015-16 for evaluation of wheat varieties under natural conditions in intermediate hill zone of Jammu region at K.V.K., Reasi, SKUAST-Jammu. The experiments were laid out in randomized block design (RBD) with three replications using a stripe rust susceptible wheat cultivar (PBW 550). The varieties were grown in lines of 1.5 m length with a row to row spacing of 22.5 cm in three replications in last week of November. Whole of the experimental area was surrounded by border rows of two lines of susceptible cultivar (PBW-343). Disease incidence and disease severity were recorded at fifteen days interval from February onwards till harvesting of the crop

An experiment on chemical management of stripe rust of wheat was conducted at Chatha farm, Jammu during three crop seasons (2013-14, 2014-15 and 2015-16). Stripe rust susceptible variety, PBW 343 was sown in field in 3 x 4 m^2 plots size with 9 treatments including control with four replications in RBD design. Stripe rust was created by artificial inoculation of mixed pathotypes spores. Fungicidal sprays were given at 15 days interval.

Results

The percent disease severity ranged from 1.0 to 15.30 per cent in $10^{th}$of February, 2014-15 while it was 1.0 to 13.39 per cent in 2015-16 but by the end of crop season the disease severity on different wheat cultivars varied from 9.4 to 62.5 per cent during 2014-15 while it was 9.5 to 62.5 per cent during 2015-16 with highest in PBW-550 (62.5 per cent) followed by HD-2967 (60.5 per cent) and it was lowest on WH-1105 (9.4 per cent) and VL-907 (9.8 per cent) as mentioned in Figure 1. From these results, it is apparent that stripe rust infection was well established in all the wheat varieties screened for the stripe rust infection.

Out of 20 wheat cultivars tested, var. PBW-343, PBW-550, HD-2967 found to be susceptible. Raj-3077, PBW-373, PBW-175, PBW-527, PBW-621, HS-490 were found to be moderately susceptible and RSP-561, Raj-3765, PBW-664, PBW-590 were found to be moderately resistant with Area Under Rust Progress Curve (AURPC) in the range of 771.75 to 849.75 during 2014-15 and it was 798.75 to 855.75 during 2015-16 and 324.0 in VL-907 during 2015-16.Wheat cultivars WH-1105, WH-1121, WH-1124, WH-1142, VL-804, VL-907, HS-240 were observed as resistant against stripe rust with Area Under Rust Progress Curve (AURPC) of during 2014-15 was in the range of 308.25 to 350.25 while during 2015-16 it was 324 to 351.75.

The stripe rust showed increased severity in all the areas as compared to the preceding year of 2014-15 to 2015-16 with maximum incidence and severity of 62.5 per cent and 60.5 per cent, respectively, Although the replacement of susceptible varieties was through on mass scale but the presence of inoculums increased the disease severity in the successive year, whereas in the intermediate hill region, the maximum area under cultivation of wheat was rainfed and the average rainfall (84.30 mm) was also high and was in consonance with the critical water requirement stages of the crop. Due to low relative humidity, the rate of disease development was slow and the hosts showed tolerance response against the disease. It revealed that the weather conditions along with cultivation of susceptible and primitive varieties, usual delayed sowing, and presence of green bridges all might have contributed to the development of disease in all the three cropping seasons (Figures 3).

Results

The result revealed that all the treatments records significant reduction of disease incidence. Two sprays of Tilt (Propiconazole) (0.01 per cent) at 15 days interval was found maximum grain yield (44.95q/ha) complete control of yellow rust. The highest yield (44.95q/ha) was found in the plot treated with two sprays of Tilt @0.1 per cent followed by two spray of Folicur (43.47 q/ha) and Bayleton @0.01 per cent (43.2q/ha) respectively. Two sprays of Tilt (0.01 per cent) at 15 days interval was minimum 4.03 per cent disease severity followed by Folicur (5.29 per cent) and Bayleton @0.01 per cent (8.90 per cent) respectively. 34.60 per cent yield increased treated with two spray of Tilt (0.01 per cent) at 15 days interval.

Recommendations

From the two years result it may be recommended to the cultivation of var. WH-1105, WH-1121, WH-1124, WH-1142, VL-804, VL-907, HS-240 in the Reasi and Udhampur districts of Jammu region. The present investigations was carried out for two consecutive seasons during Rabi 2014-15 and 2015-16 for evaluation of wheat varieties under natural in the intermediate hill zone at Reasi district. Out of

20 varieties evaluated, WH-1105, WH-1121, WH-1124, WH-1142, VL-804, VL-907, HS-240 found to be resistant; RSP-561, Raj-3765, PBW-664, PBW-590 found to be moderately resistant; Raj-3077, PBW-373, PBW-175, PBW-527, PBW-621, HS-490 found to be moderately susceptible and PBW-343, PBW-550,HD-2967 found to be susceptible. Rust progressive curve was highest 2377.42 during 2014-15 due to conducive environmental conditions while it was 2362.4 in 2015-16.

Two sprays of Tilt (Propiconazole) (0.01 per cent) at 15 days interval at diseases initiation stage are recommended for complete control of yellow rust disease.

Impact

It will enhanced wheat yield 15-35 per cent through disease control and also improved socio economic status of farmers.

Table 1: Performance of Wheat Varieties 2014-15 and 2015-16 in Rainfed Conditions in Reasi

Disease Score	*Disease Reaction*	*Germplasm*
1	Immune (0)	Nil
2	Nearly immune (1 to 5 per cent)	Nil
3	Resistant (6-10 per cent)	WH-1105, WH-1121, WH-1124, WH-1142, VL-804, VL-907, HS-240.
4	Moderately resistant (11-25 per cent)	RSP-561, Raj-3765, PBW-664, PBW-590.
5	Moderately susceptible (26-40 per cent)	Raj-3077, PBW-373, PBW-175, PBW-527, PBW-621, HS-490
6	Susceptible (41-65 per cent)	PBW-343, PBW-550,HD-2967.
7	Highly susceptible (>65 per cent)	Nil

Table 2: Performance of different Fungicides against Stripe Rust

Treatment	*Rate (per cent)*	*Disease Severity (ACI)*			*Mean Severity (ACI)*	*Grain Yield (q/ha)*			*Mean Grain Yield (q/ha)*	*Yield Increased (per cent)*
		2013-14	*2014-15*	*2015-16*		*2013-14*	*2014-15*	*2015-16*		
T1: One spray of Tilt 25 EC @ 0.1 per cent	0.01 per cent	21.25	22.74	20.85	21.60	41.67	40.15	42.62	41.48	34.60
T2: Two spray of Tilt 25 EC @ 0.1 per cent	0.01 per cent	0	5.25	6.90	4.03	44.77	43.23	44.95	44.31	38.78
T3: Two sprays of Mancozeb 75WP @ 0.25 per cent	0.25 per cent	56.25	62.33	47.12	55.22	30.50	32.2	33.30	32.00	15.22
T4: Three sprays of Mancozeb75WP @ 0.25 per cent	0.25 per cent	45.75	58.23	48.33	50.76	35.44	35.45	34.82	35.23	23.01
T5: One spray of Folicur 250 EC @ 0.1 per cent	0.01 per cent	23.75	22.73	20.20	22.21	40.77	39.84	40.81	40.47	32.97
T6: Two sprays of Folicur 250 EC @ 0.1 per cent	0.01 per cent	0.00	7.51	8.39	5.29	44.25	42.88	43.28	43.47	37.59
T7: One spray of Bayleton 25 WP @ 0.1 per cent	0.01 per cent	26.75	27.98	24.4	26.37	40.25	38.94	39.64	39.61	31.51
T8: Two sprays of Bayleton 25 WP @ 0.1 per cent	0.01 per cent	7.5	8.70	10.52	8.90	43.88	42.26	43.25	43.13	37.10
T9: Control		90.00	78.66	89.50	86.05	25.50	28.62	27.26	27.12	-

R3 T9

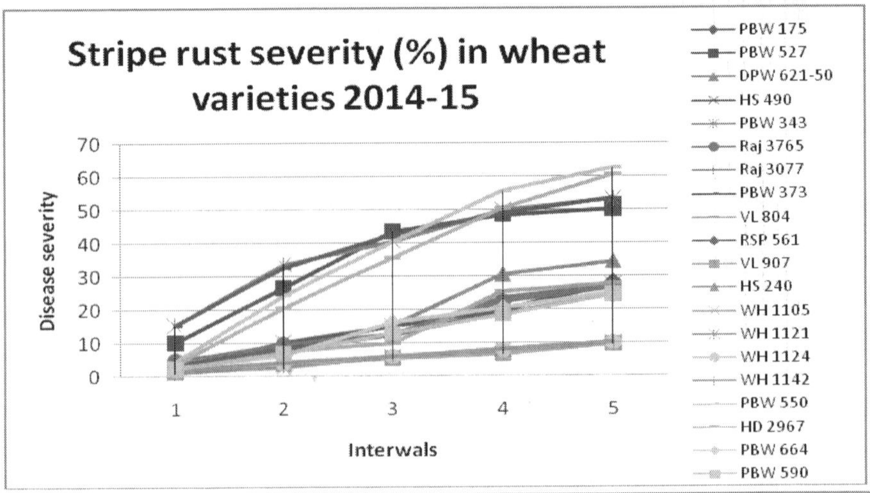
Stripe rust severity (%) in wheat varieties 2014-15
Disease severity
Interwals
PBW 175
PBW 527
DPW 621-50
HS 490
PBW 343
Raj 3765
Raj 3077
PBW 373
VL 804
RSP 561
VL 907
HS 240
WH 1105
WH 1121
WH 1124
WH 1142
PBW 550
HD 2967
PBW 664
PBW 590

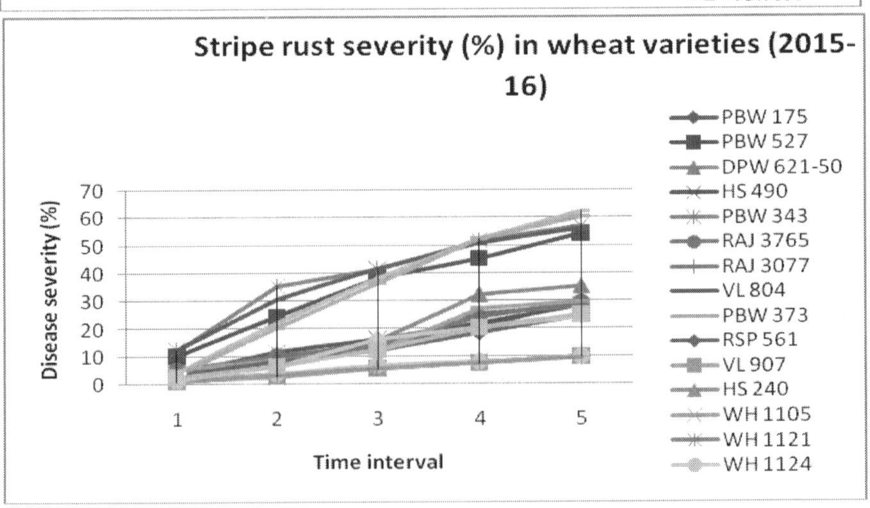
Stripe rust severity (%) in wheat varieties (2015-16)
Disease severity (%)
Time interval
PBW 175
PBW 527
DPW 621-50
HS 490
PBW 343
RAJ 3765
RAJ 3077
VL 804
PBW 373
RSP 561
VL 907
HS 240
WH 1105
WH 1121
WH 1124

71. Management of Rhizome Rot of Ginger at Farmers' Field in Jammu Region

***Contact Scientist:* Dr. Shahid Ahamad**

Rhizome rot of ginger is caused by *Pythium aphanidermatum* (Edson) Fitz and is a major constraint in the production of healthy rhizome and sometimes causing total failure of crop. It is a major disease in Reasi and Udhampur district of Jammu region. It is a rhizome as well as soil borne disease and the symptoms can be seen from July. Yellowing of leaves appear first on the lower leaves and proceeds to upper leaves. Roots arising from the affected rhizome become rotten and show brown discoloration of the rhizome tissue. Sometimes the pseudostem comes off easily with a gentle pull. The rotten parts attract other fungi, bacteria and insects. During the rainy season, this disease spreads very fast from infected field to healthy field. During the wet weather conditions, the disease causes severe leaf blight and results in significant reduction in the number and size of rhizomes thereby resulting yield losses up to 66 per cent.

Methodologies

The present research was carried out at K.V.K., Reasi on farmers' fields, during *Kharif*-2014 and *Kharif*-2015. The local ginger rhizomes JK selection were planted in field at 25 x 15 cm spacing with three replications in randomized block design. The six treatments were applied in treated plots along with control. Treatments comprises asT1-Farmers' practice (without rhizome treatment), T2 - recommended practice SKUAST-Jammu (rhizome treatment with Carbendazim + Thiram (1:1 ratio), T3 - T2 + drenching with Metalaxyl MZ 0.3 per cent, T4 - drenching with Carbendazim 0.1 per cent, T5 - drenching with Metalaxyl MZ @ 0.3 per cent and T6 - drenching with Mancozeb @ 0.2 per cent. Recommended agronomic practices were followed to raise the crop. The observations on per cent plant dried disease incidence were recorded. Efficacy of the treatments against the disease was assessed by comparing their percent plant dried, disease incidence in all replicated plots. The cost benefit ratio was also calculated for different treatments for economic viability of these treatments.

Results

Affected plants showed foliar yellowing from bottom to top and disintegration of soft rhizome tissues appeared after about 40 days after sowing. The initial symptoms were noticed on plants at about 40-45 days after sowing. The plants wilted and subsequently death of the plants was observed.

The three chemicals and five different treatments reduced disease incidence. The result observed in differential disease reaction and conclusion was drawn

Table 1: Per cent Disease Incidence and Yield of Ginger in Kharif-2014 and Kharif-2015

Treatments	*Disease Incidence (per cent)*		*Disease Incidence (per cent) Pooled*	*Yield (q/ha)*		*Pooled*	*C:B Ratio*
	2014	*2015*		*2014*	*2015*		
T1 - Check without seed treatment	35.5	55.25	45.37	90	75.0	82.5	1:2.6
T2- Seed treatment with mancozeb + carbendazim (1:1 ratio)	10.0	25.5	17.50	125	100.0	112.5	1: 3.5
T3-T2 + drenching with Metalaxyl MZ @ 0.3 per cent	7.5	10.5	9.00	140	125.0	132.5	1: 3.8
T4- drenching with Carbendazim 0.1 per cent	26.8	28.9	27.85	112	105	58.5	1:2.5
T5- drenching with Metalaxyl MZ @ 0.3 per cent	12.5	18.2	15.35	120	95	107.5	1: 3.1
T6- drenching with Mancozeb @ 0.2 per cent	18.5	17.5	18.0	118	110	114	1: 3.6
CD at 0.05 per cent	11.7	12.2					

with the two years pooled mean data. The results showed that maximum disease incidence was recorded during *Kharif*-2015 *i.e.* 55.25 per cent without rhizomes dip treatment (farmers practices in Jammu province) while lowest disease incidence was recorded during *Kharif*-2014 *i.e.* 7.5 per cent when rhizomes dip treatment was given with Mancozeb + Carbendazim (1:1 ratio) along with drenching with Metalaxyl @ 0.3 per cent followed by T2 - seed dip treatment with Mancozeb + Carbendazim (1:1 ratio). The mean disease incidence and highest yield recorded in case of treatment (T3) and maximum mean diseases incidence and minimum yield was recorded 45.37 per cent in case of treatment (T1) - check (without seed treatment). Cost benefit ratio (1:3.8) was also recorded highest during *Kharif*-2014 with rhizomes dip treatment with Mancozeb + Carbendazim (1:1 ratio) along with drenching with Carbendazim @0.1 per cent and lowest cost benefit ratio (1:2.5) was recorded 90 q/ha during *Kharif*-2014 when rhizomes was sown without dip treatment.

Infected Plants and Rhizome

72. Organic Management of Rhizome Rot of Ginger in Reasi District

Contact Scientist: **Dr. Shahid Ahamad**

Introduction

Organic farming is picking up pace in the state and there is need of awareness and trainings of farmers for organic farming. J&K has huge potential for organic farming as a large area in the state is already under semi-organic cultivation in hilly districts of the state due to the lack of availability of chemical fertilizers in these areas and the farmers of these areas hardly apply the chemical fertilizers. Rhizome rot of ginger caused by *Pythium aphanidermatum* (Edson) Fitz is a major constraint for the production of healthy rhizome, sometimes causing total failure of crop. As it is a major problems in Reasi districts, so problem was taken and trail was laid out at KVK, Reasi for organic management of the disease during 2013-14.

Methodology

Krishi Vigyan Kendra, Reasi scientists approached to the farmers of Talwara, Pouni villages of District of Reasi and motivated them for the organic farming. The KVK scientists frequently visited the village and trained the farmers for the organic farming emphasizing on organic vegetables especially ginger growing. The present research was carried out in Talwara, Reasi India during *Kharif*, 2013 and *Kharif*, 2014. The rhizomes were planted in field at 25 x 15 cm spacing with three replications in randomized block design. Seed rate used 20 q/ha. The seed rhizomes are planted at a spacing of 20-25 cm row to row and plant to plant. The best time for planting ginger in Reasi district is first fortnight of May with the receipt of pre-monsoon showers. However, under irrigated conditions, it can be planted well in advance from the middle of April. The eleven treatments were applied in treated plots along with control (water soaked/sprayed). Rhizomes treated with hot water for 2 hours upto 48°C, rhizomes are covered with polythene sheets exposed to sunlight for a period of 2 hours, Rhizome treatment with *B. diffusa* root extract @ 10 per cent, Rhizome treatment with *Trichoderma viride* @10 g/l, Rhizome treatment with Neem oil 0.3 per cent, Foliar spray of Garlic 5 per cent, Foliar spray of Turmeric 5 per cent, Foliar spray of Onion 5 per cent, Foliar spray of *B. diffusa* root extract 10 per cent, Foliar spray of Neem oil 0.3 per cent and check were maintained.

Before sowing, the rhizomes are covered with polythene sheets exposed to sunlight for a period of 2 hours for raising the temperature upto 48°C. This temperature is maintained for 30 minutes. This help to kill the bacterial and fungal pathogens present on the rhizome seeds. The rhizomes were also treated in hot water at 51°C for 10 min. All the treatments were maintained organically by adding required farm yard manure and vermin-compost and hand weeding was also done.

The roots of *B. diffusa* were cut into small pieces and allowed to dry under shade at room temperature. Root was powdered in a grinder and stored at low temperature for further use. The crude extract was prepared by making the suspension of root powder in sterile water @ 1 g/10 ml (w/v). The pulp was squeezed through two folds of cheese. The supernatant was used for seed treatment and foliar sprays. Rhizomes of ginger were soaked in *B.diffusa* root extract (10 per cent) for overnight. In control, rhizomes was soaked in plain water instead of organic agents. Three foliar sprays of 10 per cent root extracts and other botanicals, *Trichoderma* were done at weekly interval starting just after apparent symptoms. In order to cultivate ginger organically an isolation distance of 25 m wide is to be left on all around from the conventional farm. The produce from this isolated belt shall not be treated as organic. Being an annual crop, the conversion period required will be two years. Guidelines for organic production of ginger have been applied developed by Spices Board under National Programme for Organic Production (NPOP).

To make the standard plant extracts solution (100 per cent), required quantity of fresh leaf/bulb of each plant was thoroughly washed with tap water followed by distilled water and was ground @ 1 g/ml of water (1:1 w/v) and filtered through double layered cheese cloth. Spraying was done immediately after the appearance of the disease and repeated three times at 10 days interval. In each plot 25 leaves were selected randomly and per cent disease index was recorded at 70 days after sowing using 0-6 scale where 0= no symptoms; 1= 1-5 per cent leaf area covered by the pustules; 2= 5-10 per cent leaf area covered by the pustules; 3= 10-20 per cent leaf area covered by the pustules; 4= 20-35 per cent leaf area covered by the pustules; 5= 35-50 per cent leaf area covered by the pustules; 6= >50 per cent leaf area covered by the pustules were harvested by hand and rough mustard yields (kg/ha) were estimated from grain harvested in the 6 m2 plot. The data was analyzed statistically.

Results

Results indicated that the per cent mean disease incidence was found highest in the case of check *i.e.* 77. per cent while it was the highest in Kharif 2013 *i.e.* 82.5 per cent and 72.5 per cent during kharif 2014. The lowest per cent mean disease incidence was recorded in case of foliar spray of *B. diffusa* root extract 10 per cent *i.e.* 23.5 per cent followed by foliar spray of foliar spray of Garlic 5 per cent *i.e.* 36.5 per cent and neem oil 0.3 per cent *i.e.*37.5 per cent. While per cent diseases incidence was recorded foliar spray of *B. diffusa* root extract 10 per cent was found to be the best *i.e.*25.5 per cent and 21.5 per cent during Kharif 2013 and 2014 respectively, followed by foliar spray of Garlic 5 per cent. *i.e.* 35.5 per cent and 38.5 per cent respectively in both the years and in case of foliar spray of Neem oil 0.3 per cent. *i.e.* 42.5 and 32.5 per cent respectively in both the years. The mean yield of ginger was recorded in case of Foliar spray of *B. diffusa* root extract 10 per cent *i.e.* 127.5

Table 1: Effect of Organic Treatments on per cent Disease Index of Rhizome Rot Disease

	Treatments	*Kharif 2013*		*Kharif 2014*		*Mean of Two Years*	
		Per cent Disease Incidence	*Yield (q/ha)*	*Per cent Disease Incidence*	*Yield (q/ha)*	*Per cent Disease Incidence*	*Yield (q/ha)*
T1	Rhizomes treated with hot water for 2 hours upto 48°C.	42.5	96.5	40.5	98.5	41.5	97.5
T2	Rhizomes are covered with polythene sheets exposed to sunlight for a period of 2 hours.	48.5	90.5	45.5	93.5	47.0	92.5
T3	Rhizome treatment with *B. diffusa* root extract @ 10 per cent.	54,5	87.5	46.5	88.5	50.5	88.0
T4	Rhizome treatment with *Trichoderma viride* @10 g/l.	48.5	86.5	38.5	78.5	43.5	82.5
T5	Rhizome treatment with Neem oil 0.3 per cent.	49.5	87.5	45.5	86.5	47.5	87.0
T6	Foliar spray of Garlic 5 per cent.	35.5	112.5	38.5	115.5	36.5	114.0
T7	Foliar spray of Turmeric 5 per cent.	60.5	82.5	58.5	85.5	59.5	84.0
T8	Foliar spray of Onion 5 per cent	52.5	80.5	48.5	84.5	50.5	82.5
T9	Foliar spray of *B. diffusa* root extract 10 per cent.	25.5	125.0	21.5	130.5	23.5	127.5
T10	Foliar spray of Neem oil 0.3 per cent	42.5	105.0	32.5	118	37.5	111.5
T11	Control	82.5	75.5	72.5	80.5	77.0	78.0

q/ha followed by 114.0 q/ha in case of Foliar spray of Garlic 5 per cent and 111.5 q/ha in case of Foliar spray of Neem oil 0.3 per cent respectively. The yield during Kharif 2013 was found highest in case of *B. diffusa* root extract 10 per cent *i.e.* 125.0 q/ha followed by Foliar spray of Garlic 5 per cent *i.e.* 112.5 q/ha and Foliar spray of Neem oil 0.3 per cent *i.e.*105 q/ha. While in Kharif 2014 it was found 130.0, 118 and 115.5 q/ha in case of *B. diffusa* root extract, Neem oil 0.3 per cent and spray of Garlic 5 per cent respectively.

Recommendations

One of the predisposing factors for soft rot of ginger is an ill drained field. Proper drainage for cultivation ensures healthy crop. Therefore, water stagnation should be avoided. A raised bed of 30 cm height and 1m width is recommended. Rhizome dipped for overnight in the suspension of 10 per cent root extract of *B. diffusa* followed by 3 foliar sprays of the same concentration proved quite effective in the management of the disease as well as improved quality and enhanced quantity. Foliar spray of *B. diffusa* root extract 10 per cent was found to be the best *i.e.*25.5 per cent and 21.5 per cent during Kharif 2013 and 2014 respectively

Impact

Farmers are getting more price 50-60 per cent as compared to conventional farming.

E. Faculty of Veterinary Sciences and AH, RS Pura, Jammu

Division of Animal Nutrition

73. Condensed Tannins Enriched Multi-nutrient Blocks for Livestock Productivity Enhancement

Contact Scientist: **Dr. R.K. Sharma**

The deficiency of micro-nutrients in the livestock ration limits their productivity and performance. This is further aggravated by the wide-spread gastro-intestinal parasitic infection in these animals. The present technology addressed this problem by preparing functional feed blocks (multi-nutrient blocks enriched with locally available condensed tannins containing leaf meal mixture). These blocks when made available to the animal as a supplement are supposed to improve animal health and production and control gastro-intestinal nematodes. Preparation of these blocks use the widely acclaimed feed densification technique that ensures minimum wastage of nutrients; uniform intake and ease of use, storage and transportation. These blocks also offer opportunity to strategically use tanniniferous tree leaves, which otherwise remain under-utilized as alternative animal feed and/or functional feed.

Procedure of Block Preparation

The blocks were prepared by using manually operated densification machine designed and developed by Division of Animal Nutrition, FVSc and AH, SKUAST-Jammu. This machine uses the cold pressing method for densification.

Following ingredients are required for functional feed block preparation: Mustard oil cake (19.5 per cent), wheat bran (3.50 per cent), leaf meal mixture (LMM; 16.50 per cent), molasses (33.00 per cent), urea (8.50 per cent), limestone powder (6.50 per cent), dicalcium phosphate (8.00 per cent), mineral mixture (4.00 per cent) and common salt (1.50 per cent). Depending upon availability and market price, other oil cakes, cereal grains and their by-products can also be used for preparation of functional feed blocks. Urea is generally limited to 8-10 per cent and molasses is limited to 40 per cent. Limestone, dicalcium phosphate, mineral mixture and common salt are usually limited to total 10-12 per cent with about

Manual Densification Machine for Preparation of Functional Feed Block

Mixing of Ingredients and Freshly Prepared Blocks

Blocks Ready for Storage, Distribution and Use

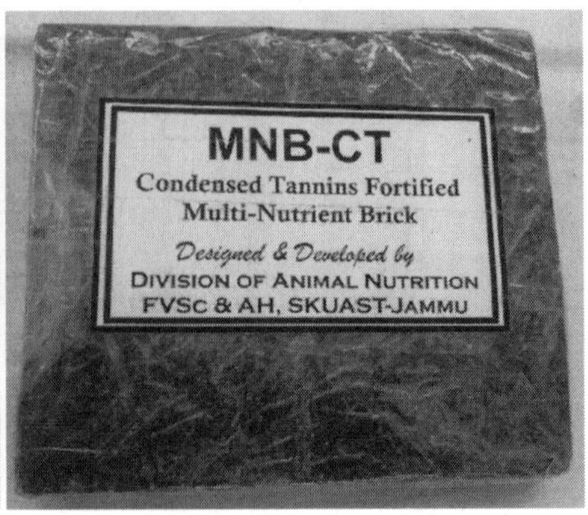

Functional Feed Block

equal amount of each supplement. Leaves of condensed tannins rich trees such as Jamun, Guava and Subabul are used for preparation of LMM. For this purpose, tree leaves are harvested in bulk and sun-dried on a pucca floor to reduce the moisture level to about 10-15 per cent. Frequent turning is required during the process of sun-drying. The dried leaves are then mixed and pulverized to powder form to prepare LMM.

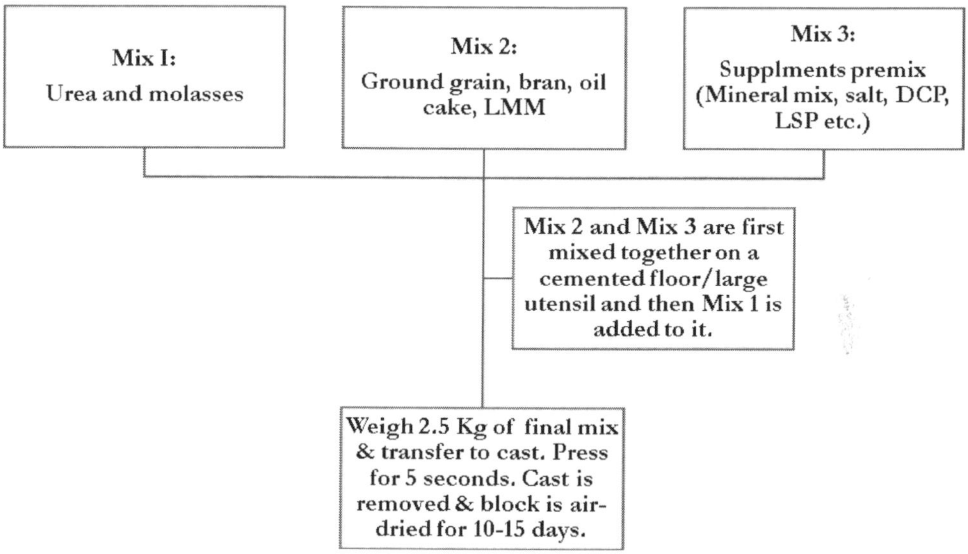

Table 1: Performance of Goats Fed Feed Blocks with or without Functional Ingredient for 75 Days

Attributes	*Groups**			*P Values*
	Negative Control	*Infected Control*	*Infected Treatment*	
Live weight change (kg)	0.70	-7.93	-3.27	-
DM digestibility (per cent)	$55.96^{b} \pm 0.54$	$52.75^{a} \pm 0.48$	$55.00^{ab} \pm 0.69$	<0.01
Nitrogen balance (g/d)	$2.57^{b} \pm 0.32$	$1.01^{a} \pm 0.26$	$2.67^{b} \pm 0.12$	<0.01
Haemoglobin (g/dl)	$10.70^{c} \pm 0.12$	$9.27^{a} \pm 0.29$	$10.28^{b} \pm 0.14$	<0.01
Cell mediated immune response (per cent) at 96 hrs	$104.03^{a} \pm 2.99$	$100.02^{a} \pm 2.27$	$132.33^{b} \pm 7.19$	<0.01
HA titre against sheep RBC (humoral immune response)	$1.656^{b} \pm 0.19$	$0.903^{a} \pm 0.12$	$3.161^{c} \pm 0.19$	<0.001
Faecal egg count for *H. contortus (no./g faeces)* ^	—	525.00±94.84	106.25±23.95	<0.05

Attributes	*Groups**			*P Values*
	Negative Control	*Infected Control*	*Infected Treatment*	
Cost of supplementation (Rs./day/animal)	0.98	1.04	0.99	-

abcMeans bearing different superscript within a row differ significantly

^Mean of fortnightly observation made through trial duration of 75 days.

*Groups

- **Negative control:** Uninfected animals supplemented with feed blocks without condensed tannins (functional ingredient).
- **Infected control:** Infected animals (dosed with 1500 *H. contortus* L_3 Larvae per animal) supplemented with feed blocks without condensed tannins (functional ingredient).
- **Infected treatment:** Infected animals (dosed with 1500 *H. contortus* L_3 Larvae per animal) supplemented with feed blocks containing condensed tannins (functional feed blocks)

Advantages of Usage of Functional Feed Blocks in Livestock Rations

- These blocks can be easily stored, offered and transported with minimum wastage.
- Feed blocks with functional ingredient (condensed tannins rich LMM) is cheaper than blocks without LMM (Rs. 12.62 vs Rs. 14.53 per kg).
- Efficient utilization of locally available feed resources
- Improvement in the nutritional status of livestock (better nutrient digestibility and nitrogen balance)
- Increase in micro-nutrients supply to the animals thereby improving production, reproduction performance and immunity
- Reduction in overall feeding and animal rearing cost (improved digestibility and less damage due to infections).
- Reduction in the use of anthelmintic drugs.
- Provision of promising alternative balanced feed resource during natural calamities.
- Can be an optimum feed resources for kandi belt of Jammu region.

Recommendation

Functional feed blocks may be supplemented *ad libitum* to ruminant livestock feeding regimen for augmenting health and production with concurrent reduction in cost of rearing.

Division of Veterinary Parasitology

74. Assessment of Economic Losses Caused by *Przhevalskiana silenus* Infestation in Goats of Jammu Region and Development of Cost Effective Treatment Regimen

Contact Scientist: **Dr. Rajesh Katoch**

Methodologies

Assessment of Economic Losses

The study was carried in Jammu province of Jammu and Kashmir state of North India that lies between $32^{o}442$ to $33^{o}532$ North latitude and 74° to $75^{o}542$ East longitude. Studies on economic losses due to hide damage and carcass trimming were carried on slaughtered animals in abattoir. These goats were brought at Municipal Committee abattoir at Jammu for slaughtering from different districts of the Jammu province. Effect on body weight gain and milk production was evaluated on goats of beetal breed, which had spent the whole summer on the same grazing area and were found naturally infested with *P. silenus* larvae by the end of February of next year. Finally the total economic loss was calculated by considering the losses from hide depreciation, meat loss (due to trimming and reduced weight gain) and decreased milk production.

Table 1: Mean Monthly Body Weight (kg) of Animals Infected with *P. silenus* in an Annual Cycle of the Infection

Group	*Mean Body Weight (kg)*													
	*Month**													
	July	*Aug*	*Sep*	*Oct*	*Nov*	*Dec*	*Jan*	*Feb*	*Mar*	*Apr*	*May*	*Jun*	*Mean ± SEM*	*P Value*
Warbled (n = 42)	20.21^{a} ± 0.14	21.24^{b} ± 0.12	22.01^{c} ± 0.10	22.75^{d} ± 0.08	23.49^{e} ± 0.12	23.90^{f} ± 0.06	24.45^{g} ± 0.07	25.03^{h} ± 0.05	25.71^{i} ± 0.05	26.61^{j} ± 0.06	27.88^{k} ± 0.07	29.12^{l} ± 0.05	24.37 ± 0.17	0.000
Non-Warbled (n = 18)	20.48^{a} ± 0.016	21.19^{b} ± 0.22	22.20^{c} ± 0.19	23.34^{d} ± 0.10	24.49^{e} ± 0.12	25.73^{f} ± 0.09	26.90^{g} ± 0.10	28.47^{h} ± 0.08	29.82^{i} ± 0.05	31.29^{j} ± 0.11	32.73^{k} ± 0.05	34.47^{l} ± 0.08	26.77 ± 0.30	0.000
Mean ± SEM	20.29^{a} ± 0.11	21.22^{b} ± 0.11	22.20^{c} ± 0.09	22.93^{d} ± 0.07	23.79^{e} ± 0.11	24.45^{f} ± 0.12	25.29^{g} ± 0.16	26.06^{h} ± 0.21	26.95^{i} ± 0.25	28.01^{j} ± 0.28	29.33^{k} ± 0.29	30.73^{l} ± 0.32	25.09 ± 0.13	0.000
P value	0.256	0.860	0.164	0.000	0.000	0.000	0.000	0.000	0.000	0.000	0.000	0.000	0.000	

Development of Cost Effective Treatment Regimen

A total of two hundred eighty goats, aged 1 to 3 years, were individually marked, weighed and then allocated according to their age into seven groups named of forty animals each. The trial was performed in mid August, with an aim to attack the first instar subcutaneous larvae. These goats had never previously been treated for GWFI. The animals in Group I, II and III were administered pour-on ivermectin preparations at 2, 5, and 200 μg/kg body weight, respectively. Groups IV, V and VI were administered pour-on eprinomectin at 25, 50, and 500 μg/kg body weight, respectively. Animals in group VII were kept as untreated control. All the treatments were administered on the same day. The backs of study animals were checked by palpation at monthly interval between December 2011 and March 2012 to detect the presence of third stage larvae of *P. silenus*

Figure 1: A: Mature Larvae of warble fly in subcutaneous tissue of native goat of Jammu; B: Trimmed carcass of affected goat.

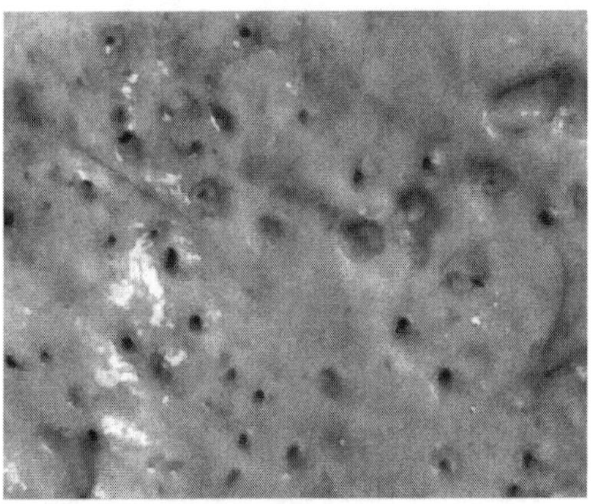

Figure 2: Holes in Goat Hide Caused by *P. silenus* Larvae.

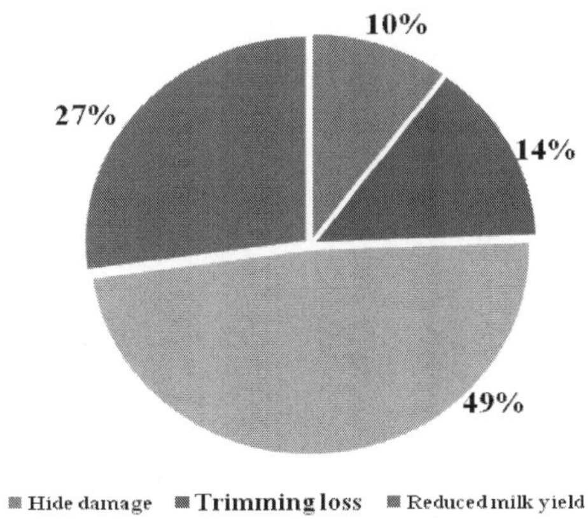

Figure 3: Various Losses Due to *P. silenus* Infestation Expressed as per cent of Total Economic Loss.

Recommendations

- In one cycle of the insect, adult *P. silenus* fly is active from April to June months, and results in losses of Rs. 73.26 million per yearin terms of decreased meat and milk production and depreciation of hide
- Application of single dose of minidose of pour on and injectable ivermectin (5µg/kg body weight) is most suited for eradication campaigns under field conditions.

Benefits/Impact

- Significance of the disease was established and highlighted to policy makers for research funding and hence a project worth Rs. 250.32 Lakhs was sanctioned by Indian Council of Agricultural Researchch- New Delhi under ICAR National Fellow Scheme.
- Cost of ivermectin treatment reduced by 1/40th times and can be used for eradication campaign

75. Detection of Acaricide Resistance in Ticks

Contact Scientist: Dr. Rajesh Katoch

- The tick infestation in domestic animals is a common and serious problem and is responsible for high economic losses.
- Among different tick species, cattle tick, *Rhipicephalus* (*Boophilus*) *microplus* is the major tick species infesting dairy animals in Jammu and Kashmir.
- It is responsible for low weight gain and milk production. It causes hide damage, skin irritation, injury to udder and anaemia (adult and larval ticks can suck 0.5-2.0 ml blood per day). It also transmits *Babesia* spp. and *Anaplasma* spp.
- The tick control is mainly focused on the use of chemical acaricides. The long term use of these acaricides has limited efficacy in reducing tick infestations due to the development of resistance. The acaricide resistance in ticks has great epidemiological significance, as the resistant population becomes more prolific and pathogenic.

Prevalence of Ticks in Jammu Region

1. Ticks are mainly found during summer and monsoon seasons *i.e.* from May to October in the plain irrigated (Jammu and Samba) and low hilly track (Poonch, Rajouri, Udhampur, and Kathua) of Jammu region.
2. In the high altitude areas (Doda and Kishtwar districts) of Jammu region, the ticks are found mainly during October to January when the animals are kept indoor.

Methodology

Collection and Preparation of Ticks

Ticks were collected from eight districts of Jammu region namely: Jammu, Samba, Rajouri, Poonch, Doda, Kishtwar, Kathua and Udhampur. The collected ticks brought to the laboratory in wide mouthed glass jars sealed with muslin cloth to allow air and moisture exchange. Then thoroughly washed with tap water and dried on filter paper, and identified. These ticks were used in the Adult Immersion Test (AIT) or were incubated (approximately 4 weeks) at a temperature of 27 ± 2°C and relative humidity of 80 ± 5 per cent to collect larvae for Larval Packet Test (LPT) for the detection of acaricide resistance.

Table 1: Acaricide Resistance Status in Jammu Region

Sl.No.	*District*	*Status of Deltamethrin*	*Status of Cypermethrin*	*Status of Fenvalerate*	*Status of Amitraz*	*Status of Diazinon*	*Status of Malathion*
1	Jammu	Severe	Moderate	Mild	Mild	Moderate	Mild
2	Samba	Severe	Moderate	Mild	Mild	Moderate	Susceptible
3	Rajouri	Severe	Low	-	Susceptible	-	Susceptible
4	Poonch	Moderate	Moderate	-	-	-	-
5	Doda	Susceptible	Susceptible	Susceptible	-	Susceptible	-
6	Kishtwar	Low	Susceptible	-	-	-	-
7	Kathua	Moderate	Low	Mild	-	Moderate	-
8	Udhampur	Moderate	Low	-	Mild	-	Mild

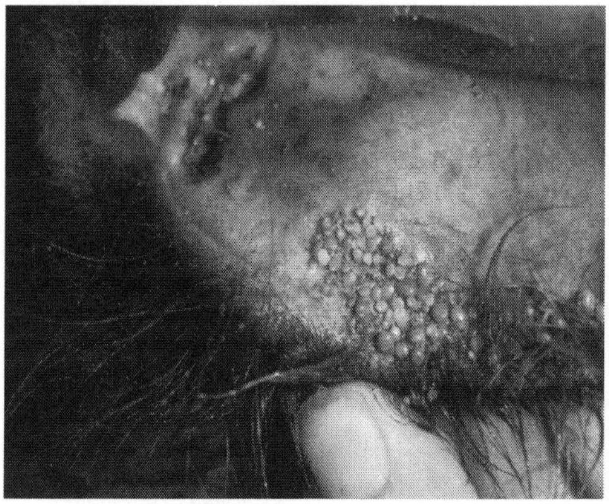

Tick Infestation in Ear

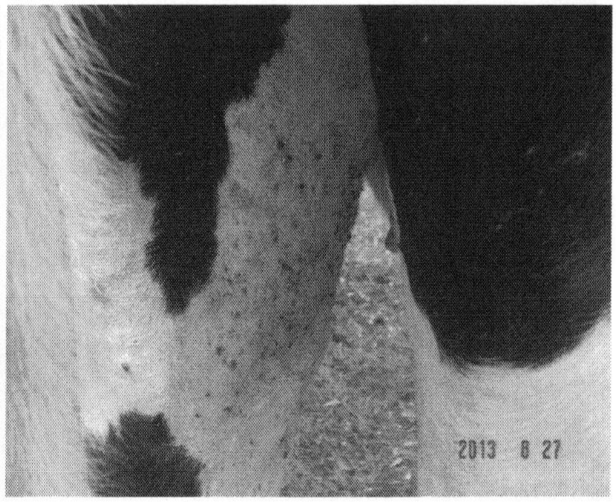

Tick Infestation in Inguinal Region

Engorged Adult Female Ticks

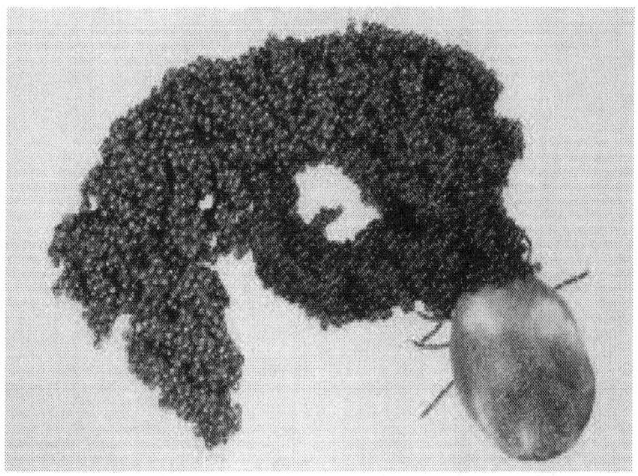

Egg-Laying of a Female Tick

Recommendations

1. Ivermectin @ 0.2 mg/kg and Amitraz @ 250 ppm
2. Poultry feed on ticks, so practicing backyard poultry will reduce the number of ticks.
3. Burning of semi-Pakka floor of animal sheds including manger (burn dry grass).
4. Filling of cracks and crevices in animal shed.
5. Spray of insecticides in the animal shed where kachha floor is practiced.
6. Burning of pastures- It is useful when different stages expected to be off the host.
7. Cultivation of land- This reduces the number of ticks due to less movement of domestic and wild animals as well as creating conditions unsuitable for tick development.

8. Uses of herbal acaricides- Leaves of Tobacco, Leaves and barks of Neem, Sitaphal seeds.
9. Chemical treatment is repeated once after 3-4 weeks post-application following other sustainable control methods.
10. Synthetic pyrethroids (Deltamethrin and Cypermethrin) and organophosphates (Malathion and Diazinon) should replace by formamidines (Amitraz) and macrocyclic lactones (Ivermectin).

Benefits/Impact

Awareness about breeding places should be given to farmers, so that the life cycle of ticks can be broken. It will provide a long term relief from tick infestation besides minimum use of the drug which is economical to farmers as well as prolongs the life of drug.

Division of Veterinary Medicine

76. Establishment of Early Diagnostic Markers for the Diagnosis of Bovine Mastitis

***Contact Scientist:* Dr. Neelesh Sharma**

Bovine Mastitis is a common disease entity of dairy cows, accompanied by physical, chemical, pathological and bacteriological changes in milk and glandular tissue. Bovine mastitis, defined as inflammation of mammary gland, can have an infectious or non-infectious etiology. The disease occurs mainly in two forms: Clinical and Subclinical mastitis. Clinical mastitis is a later stage and once developed, difficult to treat it, however which can be easily noticed by farmers through development of clinical sign such as swelling of udder/teat, blood, clots in milk. Subclinical form is most important and about 70 per cent of animals have subclinical infections. Subclinical mastitis causes heavy economic losses to dairy farmers than clinical form because it usually un-noticed by the farmers. Early and accurate diagnosis is the challenge for the farmers and veterinarians. Intramammary infections (Mastitis) may induce an acute phase response is triggered by infection, inflammation, trauma, or stress.

The diagnosis of clinical mastitis on farms relies on the observation and palpation of the cow's teats and udder and the examination of foremilk for the presence of clots, flakes or abnormal colour or consistency. Moreover, few onfarm/ field tests are also available for the diagnosis of sub-clinical mastitis, but theirs sensitivity and specificity are still not upto the mark. Sub-clinical form of mastitis is most challenging than the clinical form. Therefore, need to find out most accurate and sensitive mastitis diagnostic markers.

Methodologies

Cytokines and Early Diagnostic Markers

As mastitis is the inflammatory diseases of the mammary gland, most important pro-inflammatory or inflammatory markers such as acute phase protein, cytokines were evaluated by using specific kits (IL-1α ELISA test kit®, IL-8 ELISA test kit® and ELISA kit for Bovine Haptoglobin Assays®) in blood and milk whey for early diagnosis of mastitis. However, milk somatic cell count (SCC) was evaluated only in milk. Total 18 lactating cows were randomly selected and divided in to three groups *viz.* Group-1 (control, Negative for mastitis), Group-2 (Sub-clinical mastitis) and Group-3 (Clinical Mastitis), comprising 6 animals in each group. About 5 ml blood was collected in sterilized vials for plasma separation from jugular vein of mastitic and non-mastitic cows. Final data are presented in Mean±Standard Error (SE).

About 10 ml milk was collected in sterilized vials for whey separation from mastitic and non-mastitic cows. Milk whey was obtained after removal of fat from milk by centrifuging at 10000 RPM for 10 minutes at 4 °C. Cultural examination was considered as gold standard test for mastitis confirmation. Two by two table was used to calculate the sensitivity and specificity of early diagnostic test. Milk SCC was also used as organism specific markers.

The following diagnostic test characteristics were determined using the milk bacteriological culture result as a gold standard control.

Accuracy: [TP+TN/(TP +FP+ FN+TN)] [x100]

Sensitivity: [TP/(TP + FN)] [x100]

Specificity: [TN/(FP + TN)][x100]

Positive Predictive Value (PPV): [TP/(TP + FP)][x100]

Negative Predictive Value (NPV): [FN/(TN + TN)][x100]

Positive Likelihood Ratio (PLR): [Sensitivity/(1- Specificity)]

Negative Likelihood Ratio (NLR): [1-Sensitivity/Specificity)]

where, TP- True positive, FP-False positive, TN- True negative, FN- False negative.

Findings

In the present study, we analysed cytokines (IL-1α, IL-8) and acute phase proteins (Hp) in the healthy and mastitic cows blood serum and milk whey in relation to milk SCC (Table 1). It was found that the detectable limits of IL-1α, IL-8 and Hp were 0.59±0.12 ng/ml, 29.38±0.95 pg/ml and 65.54±16.48 μg/ml, respectively in blood serum and 0.52±0.09 ng/ml, 27.68±0.84 pg/ml and 28.62 μg/ml, respectively in milk whey of healthy control cows, while milk SCC was

119.00±10.34 (10^3/ml) (Table 1). On the other hand IL-1α, IL-8 and Hp were significantly ($p < 0.01$) increased in the both blood serum and milk whey in sub-clinical and clinical mastitic cows (Table 1). Moreover, milk SCC were non-significantly increased in sub-clinical mastitis and significantly ($p < 0.01$) increased in clinical mastitis (Table 1). In milk of cows with sub-clinical mastitis (SCC greater than 200,000, ranged from 8,65,000 to 10,98,000/ml), the mean concentrations of IL-1α, IL-8 and Hp were 2.45±0.23 ng/ml; 55.08±1.89 pg/ml and 54.78±6.01 µg/ml, respectively in the milk whey. While in clinical mastitis milk (SCC ranged from 18,98,000 to 48,76,000/ml) the mean concentrations of IL-1α, IL-8 and Hp were 4.68±0.36 ng/ml; 113.00 ±8.92 pg/ml and 118.16±6.731 µg/ml, respectively. Significant differences were observed in the all cytokines and APPS concentrations of the SCC high (>200,000 cells/ml) and low (≤200,000 cells/ml) categories (Figure 1a,b,c).

Table 1: Cytokines and Hp Level in Blood Serum and Milk Whey of Cows in Relation to Mastitis and Milk Somatic Cell Count (SCC) (Mean±SE)

Attribute	*Control Cows (n=6)*	*Sub-clinical Mastitis (n=6)*	*Clinical Mastitis (n=6)*	*P Value*
Milk SCC ($x10^3$/ml)	119.00^a±10.34	944.83^a±33.96	2984.20^b±554.77	0.001
Serum IL-1α (ng/ml)	0.59^a±0.12	2.84^b±0.03	4.68^c±0.36	0.001
Whey IL-1α (ng/ml)	0.52^a±0.09	2.45^b±0.23	4.25^c±0.40	0.001
Serum IL-8 (pg/ml)	29.38^a±0.95	68.17^b±0.94	126.24^c±7.67	0.001
Whey IL-8 (pg/ml)	27.68^a±0.84	55.08^b±1.89	113^c±8.92	0.001
Serum Hp (µg/ml)	65.54^a±16.48	486.44^b±49.40	777.16^c±49.67	0.001
Whey Hp (µg/ml)	28.62^a±3.00	54.78^b±6.01	118.16^c±6.73	0.001

a,b,c Means with different superscripts within a row differ significantly ($p < 0.01$) level.

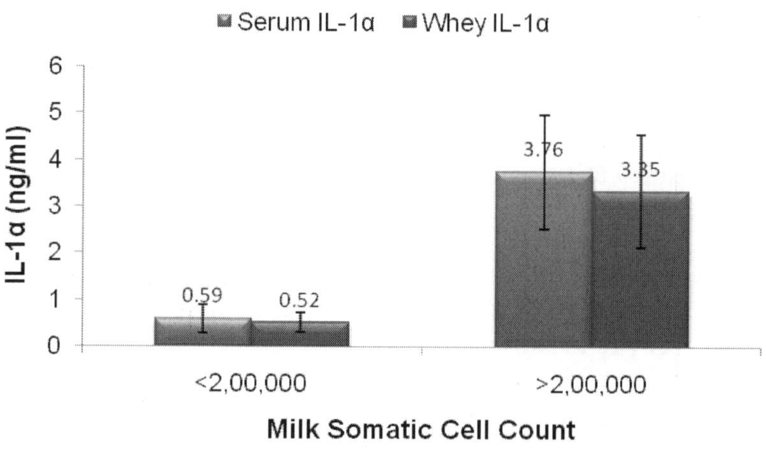

Figure 1a: Graph Showing Blood Serum and Milk Whey IL-1α Concentration

(ng/ml) in Two SCC Categories of Composite Milk Samples.

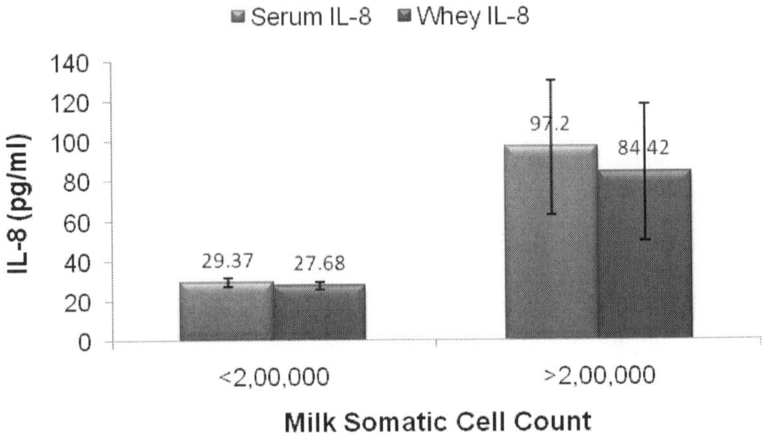

Figure 1b: Graph Showing Blood Serum and Milk Whey IL-8 Concentration (pg/ml) in Two SCC Categories of Composite Milk Samples

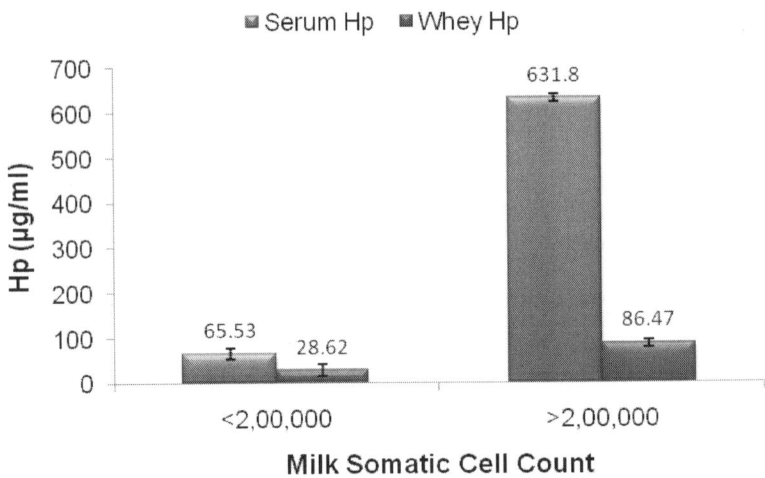

Figure 1c: Graph Showing Blood Serum and Milk Whey Hp Concentration (µg/ml) in Two SCC Categories of Composite Milk Samples.

The mechanisms by which the host recognizes bacteria at an epithelial surface are unknown. The first detectable reactions to the infection were vascular changes, which led to local edema and leakage of serum proteins into milk; a leukocytic influx followed within 2 to 4 h.

In the present study IL-1α was significantly ($p < 0.01$) increased with increasing trend of milk SCC and development of mastitis. Cytokines regulate the activity of the host's defense system in response to pathogens. IL-1, although pleiotropic in

Table 2: Agreement and Correlation of Various Tests used for the Diagnosis of Mastitis with Bacteriological (Cultural) Examination as Gold Standard Test

Test	*Sensitivity (per cent)*	*Specificity (per cent)*	*Accuracy (per cent)*	*PPV (per cent)*	*NPV (per cent)*	*PLR*	*NLR*	*k*	*r (per cent)*	*p-Value*
IL-8	100	100.00	100.00	100.00	100.00	-	0.00	1.00	100	0.00
Hp	100	100.00	100.00	100.00	100.00	-	0.00	1.00	100	0.00
IL-1α	100	85.71	94.40	91.67	100.00	7.00	0.00	0.88	100	0.00
SCC	92.31	80.00	88.90	92.31	80.00	4.62	0.10	0.72	92.30	0.002

PPV: Positive predictive value, NPV: Negative predictive value, PLR: Positive likelihood ratio, NLR: Negative likelihood ratio, k- Kappa (measure of agreement), r- Pearson's correlation, SCC: Somatic cell count, IL-8: Interleukin-8, Hp: Haptoglobin, IL-1α: Interleukin-1 alpha.

many of its effects, can activate B cells, T cells, macrophages, and other primary and secondary cells of the defense system

Present findings clearly demonstrated that IL-8 with increasing milk SCC and development of mastitis in cows increased the production of cytokines in both blood serum and milk whey. Interleukin-8 (IL-8) is an inflammatory cytokine that is produced by various cell types such as lymphocytes, neutrophils, monocytes, macrophages, and epithelial cells.

Hp level was also significantly increased in the mastitic cows (Table 1), which is an indicator of inflammation. In our study we also found significantly ($p < 0.01$) increased concentration of Hp (sub-clinical mastitis 54.78± 6.01 μg/ml and clinical mastitis (118.16±6.73 μg/ml) in milk whey of mastitis cows than healthy buffaloes (Table 1).

Sensitivity and Specificity of Cytokines as early Diagnostic Markers

The percentage accuracy of IL-8, Hp and IL-1α were 100 per cent, 100 per cent and 94.40 per cent, while SCC showed 88.90 per cent (Table 2). In the present study it was appears that IL-8 and Hp are the most reliable tests and similar to the bacteriological results indicating the presence of inflammatory conditions in the udder. Sensitivity and specificity of IL-8, Hp and IL-1α were 100 per cent, 100 per cent, 100 per cent and 100 per cent, 100 per cent, 85.71 per cent, respectively, while SCC had 92.31 per cent sensitivity and 80.00 per cent specificity. IL-8, Hp and IL-1α had 100 per cent correlation, whereas, SCC showed only 92.30 per cent correlation (Table 2).

Recommendations

Present study concluded and recommended that IL-8 and Hp are best early mastitis diagnostic markers/tests for mastitis in dairy animals than IL-1α and SCC. However, milk culture is gold diagnostic test but milk culture will identify the presence of mastitis pathogens only but will not provide a measure of the degree of inflammation associated with the infection.

Benefits/Impact

Early, rapid and accurate diagnosis of bovine mastitis has the potential to improve the efficacy of treatment and bacteriological cure rate, which is still challenging to vets for timely and specific treatment. Timely and early treatment of mastitis can reduce the economic losses to dairy farmers in term of cost of medicine, milk production, management of affected animals for long duration *etc.*

Acute form of mastitis has direct impact on animal welfare due to severe pain and sometimes it causes lameness. The early diagnosis may help in the timely treatment which is an important aspect for animal welfare, which is usually ignored in case of mastitis.

77. Development of Cost Effective Technology for the Management of Mastitis in Dairy Animals

Contact Scientist: Dr. Neelesh Sharma

Mastitis in dairy animals particularly in cattle and buffaloes is causing heavy economic losses worldwide including India. As India is standing at first position in the milk production but mastitis is severely affecting the milk quality and production. Mastitis is mainly a managemental diseases, and it suggested that "Prevention are better than cure" in case of mastitis. To achieve this goal, it is primary concern of dairy producers is to implement and maintain a good herd hygiene and mastitis control program. Despite the widespread of such control programs among dairy herds (teat dipping, dry cow therapy and other managemental practices), mastitis cases still to constitute a repetitive problem. The treatment of mastitis by using antibiotics is very limited (about 50 per cent recoveries). Hence, better way to control this disease by diagnosing at an early stage and applying preventive measures such as teat dip. Therefore, present study was planned to standardize the cost-effective preventive measures by using 1 per cent Potassium permanganet (KMnO4) as teat dip and udder washing solution before and after milking.

Methodologies

Study Area

The present study was conducted in the villages of R.S. Pura subdistrict of Jammu region.

Technical Plan

Total 12 lactating dairy cows were randomly selected and divided into two groups, *viz.* Group-1 as control without teat dip and Group-2 subjected for regular post-milking teat dipping with 1 per cent Potassium permanganate ($KMnO_4$) solution for the duration of 60 days. To study the effect of teat dip as preventive measure, milk samples were collected at 0 day, 15th day, 30th day, 60th day of teat dipping and 30th day post teat dipping. Milk samples were subjected for Modified California Mastitis Test (mCMT) and analysed for milk Somatic Cell Count (SCC), fat, lactose, protein, ash and solid not fat (SNF) by using Lactoscan Milk Analyser (Milkotronic Ltd., Bulgaria).

Findings

In present study, it was found that use of 1 per cent potassium permanganate reduced the number of infected quarters from 17 on day 0 to 12 upto the completion of teat dipping on day 60, while numbers of infected quarters were gradually increased in the control group (Figure 1; Table 1). On day 0, the mean SCC was

Table 1: Effect of Teat Dipping with 1 per cent Potassium Permanganate on Mastitis Status and Milk Quality of Dairy Cows (Mean±SD)

Groups	*Days of Sampling*	*No. of mCMT Positive Quarters*	*SCC ($x10^3$/ml)*	*Fat (per cent)*	*Lactose (per cent)*	*Protein (per cent)*	*Ash (per cent)*	*SNF (per cent)*
Group-1	0	16	729.33^{Aa}±176.06	3.43^{Aa}±0.24	4.82^{Aa}±0.05	3.59^{Aa}±0.18	0.75^{Aa}±0.03	8.84^{Aa}±0.30
	15th	16	1238.20^{Aa}±292.55	3.37^{Aa}±0.20	4.73^{Aa}±0.06	3.52^{Aa}±0.12	0.74^{ABa}±0.04	8.66^{Aa}±0.33
	30th	18	2538.80^{Ba}±544.10	3.18^{ABa}±0.12	4.55^{Aa}±0.08	3.42^{Aa}±0.12	0.72^{Ba}±0.04	8.59^{Aa}±0.44
	60th	20	3402.20^{Ca}±509.16	3.02^{Ba}±0.12	4.53^{Ba}±0.12	3.42^{ABa}±0.07	0.70^{Ba}±0.03	8.45^{Aa}±0.44
	90th	20	3863.51^{Ca}±754.71	2.92^{Ba}±0.26	4.60^{Ba}±0.10	3.41^{Ba}±0.08	0.67^{Ba}±0.03	8.19^{Aa}±0.64
Group-2	0	17	957.50^{Ab}±174.52	3.45^{Ab}±0.15	4.71^{Ab}±0.13	3.58^{Ab}±0.09	0.71^{Ab}±0.03	8.71^{Ab}±0.36
	15th	16	816.67^{Ab}±138.32	3.53^{Ab}±0.14	4.70^{Bb}±0.13	3.67^{Ab}±0.61	0.73^{ABb}±.02	8.89^{Ab}±0.25
	30th	14	491.00^{Bb}±185.70	3.82^{Bb}±0.16	4.72^{Bb}±0.09	3.83^{Ab}±0.87	0.73^{ABb}±0.03	8.88^{Ab}±0.20
	60th	12	208.17^{Cb}±64.53	3.90^{Bb}±0.18	3.93^{Bb}±0.06	3.68^{Bb}±0.78	0.75^{BCb}±0.04	9.00^{Ab}±0.15
	90th	10	191.50^{Cb}±46.09	3.88^{Bb}±0.19	4.84^{Cb}±0.08	3.94^{Cb}±0.82	0.77^{Cb}±0.02	9.01^{Ab}±0.23

Group-1: Control without teat dipping.; Group-2: Post-milking teat dip with 1 per cent Potassium permanganate. Teat dipping with Potassium permanganate was done upto 60 days.

A,B means with different superscripts within columns differs significantly ($p < 0.05$) within group at different time intervals.

a,b means with different superscripts within columns differs significantly ($p < 0.05$) between the groups.

higher than normal (≤2 $x10^3$/ml) in the both groups. Teat dipping effectively controlled somatic cell counts of milk from day 0 (957.50±174.52 $x10^3$/ml) to day 60 (208.17±44.53 $x10^3$/ml) in cows, which was significantly ($p < 0.05$) reduced and reached under normal range (191.50±46.09 $x10^3$/ml) after 30 days of completion of teat dipping. Whereas, in control cows milk SCC was constantly increasing from day 0 (729.33±176.06 $x10^3$/ml) to day 60 (3402.20±509.16 $x10^3$/ml) (Table 1). Present study also revealed that there was no negative impact of teat dipping on milk composition, moreover, milk constituents such as fat, lactose and protein were increased significantly after day 30 onwards in the teat dipping group in comparison to control group (Table 1).

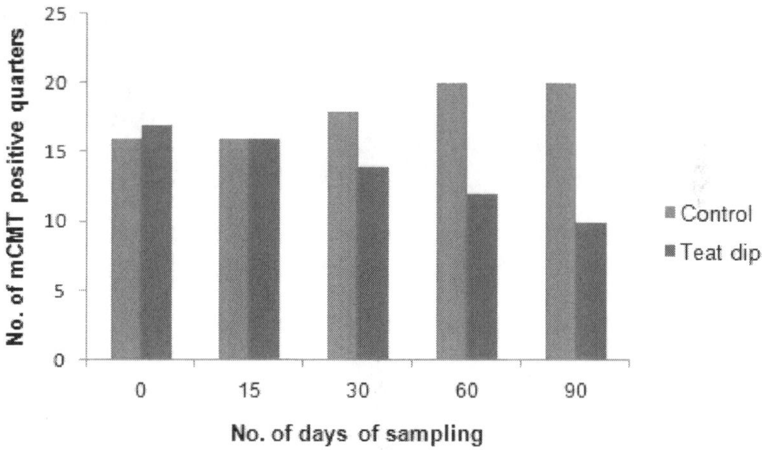

Figure 1: Effect of 1 per cent Potassium Permanganate on the Number of Mastitis Positive Quarters.

The contagious mastitis organisms are usually present on the udder and teat and they get an entry the teat canal during or immediate after milking and develop the intramammary infection. Hence, it was suggested to farmers to use the potassium permanaganet as pre-milking udder wash (Figure 2).

Recommendations

The present study recommended that the new udder infections (Mastitis) in dairy animals can be checked appreciably by following teat dipping with 1 per cent potassium permanganate solution as post-milking teat dip. Farmers were also advised for proper pre-milking washing of udder with clean water to control contamination of milk with pathogenic bacteria.

Benefits/Impact

As we know mastitis is caused by more than 137 different organisms and many of them have zoonotic importance, which may cause disease/infection such as T.B., Sore throat, diarrhea in children's *etc.* and significantly reduce the quality of milk and milk products. Hence, control of mastitis by regular use of teat dipping by using 1 per cent potassium permanganate, will have direct impact on milk quality, human health and reduce the economic losses to the farmers due to mastitis. As potassium permanganate ($KMnO_4$) is a cheapest and easily available in the market for farmers. After disseminating/transferring this technologies in the field through door stem trainings, more than 60 per cent farmers had started to adopt this technology on regular basis.

Figure 2: Preparation of 1 per cent Potassium Permanganate Solution and Udder Washing with $KMnO_4$ before Milking.

Division of Veterinary Gynaecology and Obstetrics

78. Embryo Transfer Technology in Sheep

Contact Scientist: **Dr. Anil Kumar Pandey**

Methodologies

Steps involved in the ETT:

1. Selection of donor and recipients.
2. Synchronization of estrus.
3. Superovulation of donors.
4. Detection of estrus in donor and recipients.
5. AI/natural service of donor sheep.
6. Embryo collection and evaluation.
7. Embryo transfer in recipient ewes.
8. Cryopreservation of embryos.

Criteria for Selection of Donor

Donor ewes are selected on the criteria of its genetic superiority and superior individual performance. It should have normal reproductive tract. Comparatively younger and healthy donors are preferred. Donor must have lambed at least 30 days back. For repeated Superovulation donor should be selected at least 60 days after the first superovulation treatment.

Criteria for Selection of Recipients

Recipients should be healthy and have regular reproductive cyclicity, without any abnormalities in the genital tract.

Synchronization of Estrus

Synchronization of estrus saves labour and is a key component in Artificial Insemination (AI) and Embryo Transfer (ET) programs. There are two main approaches for estrus synchronization:

Progesterone based approach: Administration of exogenous progesterone exerts a negative feedback effect over the hypothalamus and pituitary and blocks the release of natural gonadotrophins (FSH and LH). Withdrawal of progesterone block, releases larger quantities of gonadotrophins which ensure growth and maturation of follicles and onset of estrus. Most of the ewes exhibit estrus within 24-72 hrs following withdrawal of the progesterone treatment. Comercially available progesterone implants are CIDR (controlled internal drug release) for sheep, norgestomet ear implant (use 1/3rd implant available for cattle use, without estrogen and progesterone injection). A very cost effective progesterone impregnated sponge may be procured from CSWRI (ICAR), Avikanagar, Rajasthan. Progesterone implants are kept in situ for 12 days.

Prostaglandin $F_2\alpha$ based approach: $PGF_2\alpha$ terminates the luteal phase through regression of the CL in cyclic ewes. Because not all stages of the estrous cycle are receptive to the treatment, a double injection system 9-12 days apart is the most widely used approach in the ewes and the ewes should be bred at 2 to 3 days after second injection of $PGF_2\alpha$. In our embryo transfer program, we have used two doses of $PGF_2\alpha$ @10.0 mg each 10 days apart in donor and @7.5 mg in recipient ewes for synchronization of estrus.

Detection of Estrus in Donors and Recipients

Estrus detection is performed after 6-72 hrs after removal of Progesterone sponge/last prostaglandin injection. Use of appronized ram is more practical for estrus detection. Other methods are use of vasectomized male, androgenized females and with a ram after tying a string around its prepuce.

Superovulation of Donors

Superovulation among donor ewes can be induced with the use of gonadotrophins. In general gonadotrophins are administered in multiple dose injections in the last 3-4 days of synchronization treatment. Some commercial preparations of FSH available for superovulation are Ovagen and Embryo-S from ovine origin and FSH-P, Folltropin-V and Super-O from porcine origin. We have effectively used a total of 80 mg FSH (Folltropin-V) in three days tapering dose schedule (20mg, 20mg, 15mg, 15mg, 5mg and 5mg) along with eCG (Folligon) 200IU on first day of FSH.

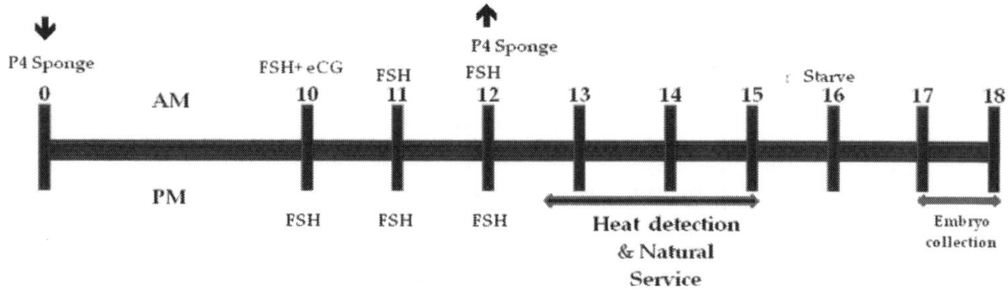

Figure 1: Schedule of Synchronisation and Superovulation Treatment of Donor Ewes.

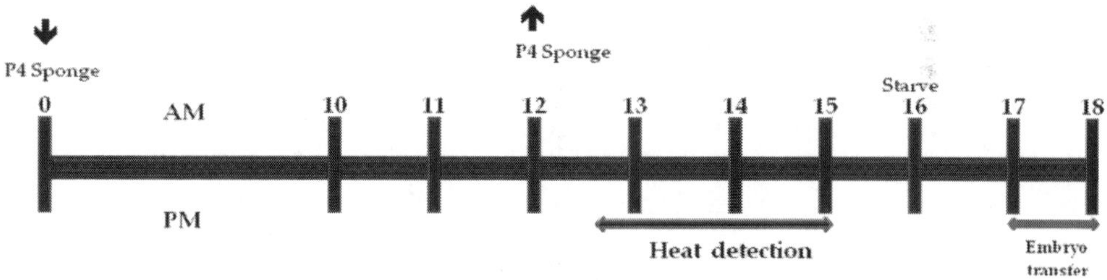

Figure 2: Schedule of Synchronisation Treatment of Recipient Ewes.

AI/Natural Service of Donor Sheep

Donors detected in heat are allowed for natural mating by the breeding ram or it may be inseminated using fresh diluted semen or frozen thawed semen. It is better to inseminate ewes on heat detection and ram may be allowed in the night hours with the donor ewes for natural service. AI with the liquid semen is performed using glass speculum and pipette and 0.1 ml of diluted semen is deposited on the mouth of the cervix. Laparoscopic intrauterine insemination can also be performed which results in high fertilization rate.

Viewing of Ovaries for Superovulatory Response with the Aid of Laparoscope

The ovaries are examined for superovulatory response on day 4 to 6 post estrus with the aid of a laparoscope. Utmost care is taken in handling the ovaries as any injury to corpus luteum leads to profuse bleeding.

Embryo Collection and Evaluation

For embryo recovery surgical mid ventral laparotomy is performed between 4 to 6 days after mating. Ewes are fasted for at least 24 hours prior to laparoscopy and/or laparotomy. The abdominal area anterior to udder is surgically prepared and the ewes are sedated with inj. Xylazine hydrochloride @ 0.29 mg/kg b.w. and locally anaesthetized by infiltration of inj. Lignocaine hydrochloride 2 per cent (5ml). Before laparotomy ewes are subjected to ovarian examination with the aid of laparoscope to determine whether they are worth attempting for surgical embryo collection. The ewes with more than 3 corpora lutea (CL) are marked as having a superovulatory response.

For embryo collection, retrograde flushing of each uterus and fallopian tube is done with a flushing media (50-60 ml, Dulbecco's phosphate buffer saline (DPBS) of pH 7.5 ± 0.3 at room temperature supplemented with 0.2 per cent bovine serum albumin (BSA), Sigma Aldrich Co., St. Louis, MO; U.S.A.) introduced near the uterine bifurcation using 18G IV cannula and collected from fallopian end catheterized by a feeding tube (size 7) in sterile 100 ml beaker from the ovarian end. Commercially available uterine flushing media such as Euroflush, IMV technologies, France are equally effective for uterus flushing and embryo recovery.

Embryo Searching, Grading and Evaluation

Flushed fluid from each uterine horns are transferred separately into sterile 90 mm petridishes for searching, evaluation and gradation of embryos under Stereo zoom microscope (Olympus, Japan). Upon location of an embryo, it is lifted with the help of an embryo lifter and transferred into small tissue culture 35 mm petridish containing approximately 1 to 2 ml of holding medium and graded as follows.

Morphological Classification of Embryos

Excellent

Embryos are perfectly symmetrical showing even granulation with well-defined distinct outline and no blastomere extrusion.

Good

Embryos showing well defined distinct outline with even granulation, some blastomere extrusion and slight degeneration; occasionally slightly asymmetrical in shape.

Fair

Embryos integral but with a hazy outline in parts; some blastomere extrusion and degeneration; occasionally asymmetrical in shape or with debris attached to zona.

Poor

Embryo showing uneven granulation and with a hazy outline; much blastomere extrusion and degeneration; sometime abnormally shaped.

Unfertilized

Granular appearance of cytoplasm, no indication of cleavage and non formation of blastomere cells. * ***Excellent, good and fair quality embryos are considered as transferable***

Laparoscope Aided Embryo Transfer

Recipient ewes are withheld for water and feed for at least 24 hours before the Laparoscopy. Ewe is restrained in dorsal recumbency on a specially designed lap cradle. All the four legs are firmly secured in the cradle and a surgical area (10-12 cm of lower belly area near udder) is prepared. Local anaesthetic, 3-4 ml of 2 per cent lignocaine hydrochloride is injected subcutaneously, 5-7 cm anterior to udder, 3-4 cm each side of the mid ventral line. Sedation in sheep (about 30 Kg Body weight) is accomplished with i/m injection of xylazine @0.29 mg per kg body weight. Animal is anesthetized by 1.5 to 2 ml Ketamine (150 -200 mg) IV. Endoscope is sterilized by immersing telescope rod in 70 per cent alcohol followed by normal saline solution during use.

Viewing of Uterine Horns

Two small (2 cm) incisions are made with angular blade in the skin, approximately 4-6 inches in front of the udder, one on each side of midline avoiding blood vessels (All surgical or laparoscopic tools are soaked in alcohol or sterilizing solution to disinfect between surgeries). Trocar and cannula for the laparoscope are inserted on the left side of midline by sliding it under the skin about one inch before inserting it into the peritoneal cavity, avoiding puncturing blood vessels or other structures. Trocar is removed as soon as it enters in the peritoneal cavity. For better visualization of internal organs in fatty animals, pneumoperitoneum apparatus can be used though for routine embryo transfer it is not essential.

Laparoscope is inserted through cannula and a manipulation probe (a knitting needles may be used by bending its 1 inch tip to 30 $^{\circ}$ angle) is inserted through second incision. With the manipulating probe it is easier to bring the uterine horn under field of vision. Uterine horns are located without observing or disturbing ovaries. Specially designed atraumatic Allis forceps are used to hold and pick out loop of the uterine horn for embryo transfer in recipients.

Embryo Transfer

Embryos are loaded either in tomcat catheter or modified aspic in between air spaces. The catheter or aspic is attached to 1 ml syringe. Recipient ovaries are viewed through laparoscope and uterine horn ipsilateral to corpus luteum is gently lifted to the surface of skin and held between thumb and forefinger. An aperture is made in the uterine horn with a small conical trocar and embryo loaded catheter is inserted in to lumen followed by embryo deposition. If Aspic is used, its needle is inserted into lumen of uterine horn and embryos are deposited.

Suturing of Incision and Post Operative Care

Surgical wound is closed with sutures. Povidone iodine ointment is applied at the suturing sites. Antibiotic injection Streptopenicillin (Dicrysticin; Sarabhai Zydus, India) @ 500 mg I/m is given just prior to laparoscopy and is continued for 3-5 days. NASID inj. Meloxicam 5mg/ml (Melonex 3ml/sheep) is given for 3 days and sutures are removed after 8-10 days.

Pregnancy Diagnosis

Pregnancy in recipient ewes are diagnosed on day 35-40 post embryo transfer with help of ultrasound scanner using 5 MHz trans-abdominal probe.

Cryopreservation of Embryos

Only good and fair quality embryos are selected for cryopreservation. Embryos are lifted out from the culture dish and washed in modified PBS with 0.3 BSA at room temperature (25°C). After washing embryos are equilibrated for 10 min in each of cryoprotectant solution containing 2, 4, 6, 8, and 10 percent glycerol in modified PBS with BSA in the ascending order of concentration. After equilibration, each embryo is loaded in the middle of a 0.25 ml embryo straw separated by two air bubbles on both sides. Date, number, species and source of embryos are mentioned on each straw for identification.

Freezing of Embryo

The loaded straws are frozen using a programmable embryo freezer and stored in liquid nitrogen container.

Thawing of Embryo

For thawing straws are placed into 20°C water bath. After thawing straws are wiped dry with tissue paper. Then the lab seal end is cut. By keeping cut end over a petridish, the factory seal end is also cut and embryo is drained into the petridish along with the medium. Then embryos are place in modified PBS with BSA in the descending order containing 10, 8, 6, 4, 2 and 0 per cent glycerol for 10 min in each concentration. Morphology of embryo is evaluated and then loaded in the straw/ tomcat catheter for transfer.

Success of Embryo Transfer

Success of ETT depends on management practices employed, method of estrus synchronization and superovulation response of donors. It also depends on breeding method used, efficiency of embryo collection technique, expertise in evaluation and transfer of embryos and various other factors affecting their survival during gestation length.

Table 1: Summary of Embryo Transfer work Done

1.	Number of donors super ovulated	43
2.	Number of donor came in heat	30
3.	Number of donor operated and flushed	27
4.	Number of embryo recovered	53
5.	Number of embryo transferred	31
6.	Number of embryo frozen and stored	14
7.	Number of recipients synchronised	76
8.	Number of recipients came in heat	52
9.	Number of recipients used for Embryo Transfer	19
10.	Number of sheep conceived through ETT	9
11.	Conception per cent of recipients	47.37 per cent
12.	Lambing per cent	66.66 per cent
13.	Live birth of lambs	8
14.	Surviving lambs	7

Table 2: Details of Lambs Born through Embryo Transfer Technology (ETT)

Sl.No.	*Tag. No./Brand No.*	*Breed*	*Date of Birth*	*Sex*
1	7ETR-14-16	Farmbred Rambouillet	14-02-2014	Male
2	7ETR-14-30	Farmbred Rambouillet	14-02-2014	Female
3	7ETR-14-17	Farmbred Rambouillet	20-03-2014	Male
4	Paint No. E4B	Farmbred Rambouillet	20-03-2014	Female
5	00619	75 per cent Dorper	24-10-2014	Male
6	00645	75 per cent Dorper	27-10-2014	Female
7	0062	100 per cent Rambouillet	02-02-2015	Male
8	0056	100 per cent Rambouillet	04-02-2015	Male

Application of Progesterone Sponge for Synchronisation Treatment.

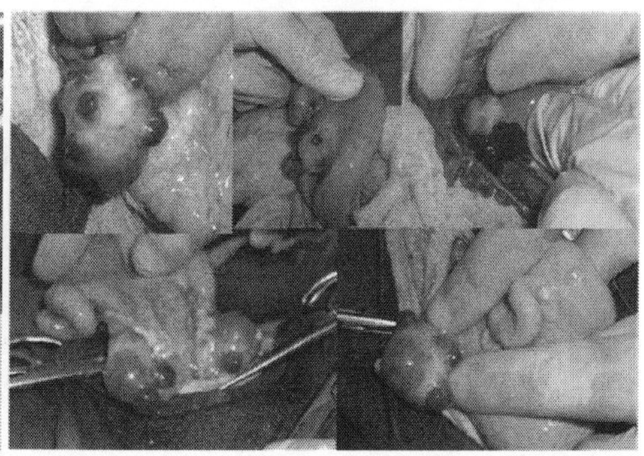

Superovulated Ovaries Showing Multiple CL and Anovulatory Follicles.

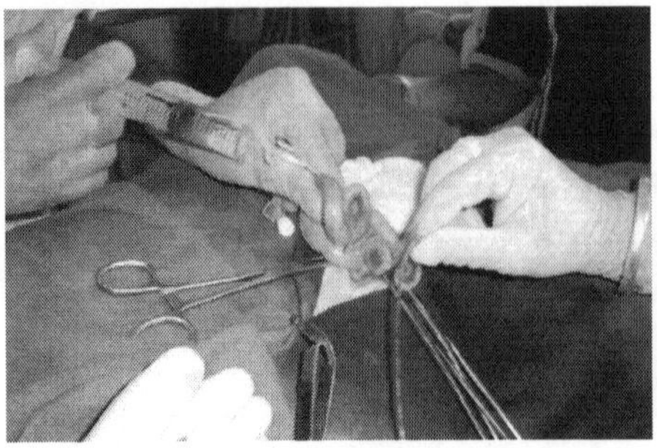

Flushing of Uterine Horns through Fallopian Tube (Retrograde flushing method).

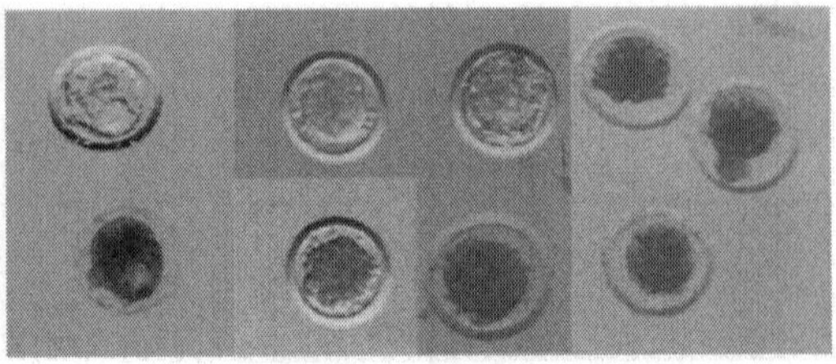

Transferable Quality Embryos (Compact morula and early blastocyst stage).

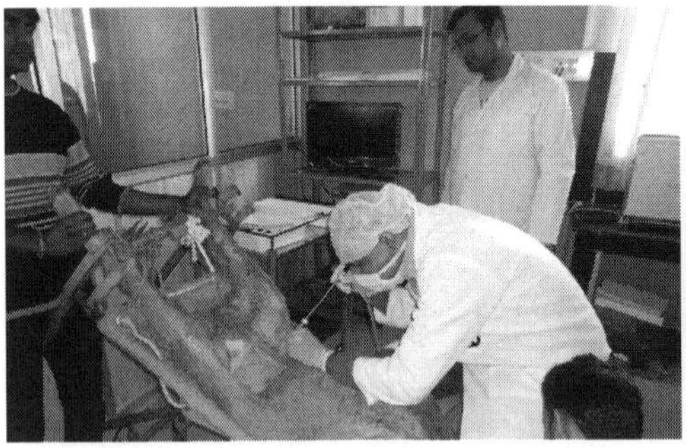

Laparoscopic Embryo Transfer: Visualization of Uterine Horns and Ovaries.

Surrogate Ewe with ETT Born Lambs (one day old).

Recommendations

Embryo transfer technology is one of the key reproductive technologies applied for enhancing genetic potential of low producing sheep and faster multiplication of superior sheep germplasm. Multiple Ovulation and Embryo Transfer enables production of large number of elite animals within a short period of time.

Benefits/Impact

1. Production of more offspring from genetically superior females.
2. Shortening generation interval by using prepubertal donors.
3. More twinning percentage by transferring bisected embryos.
4. We can obtain lambs from old, sub-fertile quality ewes.
5. Freezing of elite sheep embryos for posterity.

79. Soya Based Extender and Use of Various Semen Additives for Cryopreservation of Cattle and Buffalo Semen

Contact Scientist: **Dr. W.A.A. Razzque**

Methodologies

Soya Based Semen Extender

Routinely used semen extender contains up to 20 percent egg yolk and hence poses higher the risk of xenobiotic contamination. Egg yolk may be a prospective source of virus infections or allergic reactions. Presence of abnormal micro flora in the semen beyond certain levels may reduce the survivability of spermatozoa due to endotoxin production and also reduce fertility of cows. Presence of yolk globules interferes with the microscopic examination of diluted semen and it increases viscosity of extender. So, the presence of particulate debris in extender could be the cause of reduced fertility. Also semen extenders containing animal product components such as egg yolk and skim milk are difficult to standardize. Glycerol and several low-density lipoproteins from egg yolk are the usually used cryoprotectants in cryopreservation of bull semen since many years. But in the last few year new diluents with lecithin based cryoprotective components have been introduced into practice. Soybean contains Lecithin, a high molecular weight lipoprotein substitute for egg yolk, to prevent or repair damage to the spermatozoal plasma membrane during cryopreservation. Soya based extender is economically efficient plant based diluent for cryo preservation of bovine semen.

Use of Various Semen Additives for Cryopreservation of Cattle and Buffalo Semen

It has been observed that cryopreservation of semen is associated with oxidative stress induced by free radicals. Sperm cells have a high content of unsaturated fatty acids in their membrane but lack in a significant cytoplasmic component containing antioxidants. Hence, they are highly susceptible to lipid peroxidation (LPO) by oxygen free radicals and hydrogen peroxide (H_2O_2) which causes damage to the membrane of sperm during freezing and thawing. So, a variety of additives like antioxidants, membrane stabilizers, motility enhancers and chelating agents have been used by the scientists to protect spermatozoa from deleterious effects of cryopreservation and for improving freezability and fertility of bull semen.

Enzymatic antioxidants, also known as natural antioxidants, neutralize excess reactive oxygen species (ROS) and prevent it from damaging the cellular structure. Enzymatic antioxidants are composed of superoxide dismutase (SOD), catalase, glutathione peroxidase (GPx), and glutathione reductase (GR). These cause reduction of hydrogen peroxides to water and alcohol. SOD spontaneously dismutase superoxide anion (O^{-2}) to form O_2 and H_2O_2, while catalase converts H_2O_2 to O_2 and H_2O as shown in following equation.

$$2(O^{-}{}_{2}{}^{\cdot}) + 2H^{+} \xrightarrow{SOD} H_2O_2 + O_2,$$

$$H_2O_2 \xrightarrow{Catalase} H_2O + ½\, O_2$$

SOD protects spermatozoa against spontaneous O_2 toxicity and lipid peroxidation.

Preparation of Soya Milk

For preparation of Soya milk extender first soya milk is prepared and then it was mixed in tris buffer to make soya milk extender. The method of making soya milk comprised of heat treatment to 50 gm of soy seed in microwave oven till it become golden brown and then soaking the soy seed overnight in 0.3 percent sodium bicarbonate. Next morning, dehulled and washed with distilled water. Then boil dehulled seeds in fresh 0.3 percent sodium bicarbonate for 60 minutes. After boiling, grind seeds using a homogeniser and then prepare slurry with boiling water (1:8). Filter the slurry through four folds of muslin cloth to obtain soymilk.

Preparation of Soya based extenders (SBE): For preparation of Soya milk extender first soya milk is prepared as per given method and then it was mixed in tris buffer to make soya milk extender as per following composition.

Composition of Soya Based Dilutor for Cryopreservation of Cattle Semen

Tris buffer (modified)*	71 parts
Soymilk	29 part
Glycerol (v/v)	6.9 percent for cattle

Composition of Soya Based Dilutor for Cryopreservation of Buffalo Semen

Tris buffer (modified)*	68.6 parts
Soymilk	25 parts
Glycerol (v/v)	6.4 percent for cattle

***Composition of Tris Buffer (Modified)**

Tris (hydroxy methyl amino methane)(AR) =	2.42 gm
Citric acid (AR)	1.38 gm
Fructose (AR)	1 gm
Glycerol	6.4 ml
Penicillin G sodiumStreptomycin sulphate	1000 IU1000 µg

Preparation of SBE-CY Dilutor

2.0 mM (0.35126 mg/ml) of L-Cysteine hydrochloride (HiMedia Laboratories Pvt. Ltd., Mumbai) was added to per ml of soya milk extender.

Preparation of SBE-G Diluter

5.0 mM (1.53 mg/ml) of L-Glutathione (Sigma-Aldrich Pvt. Ltd., USA) was added to per ml of soya milk extender.

SOD and CAT Additive

For cryopreservation of buffalo bull semen, Superoxide dismutase and catlase is added in soya based extender each @100 units/ml.

Dilution of Semen

After initial evaluation, semen sample are diluted in such a way that it contained 80 million spermatozoa/ml post dilution for packing in 0.25ml French mini straws.

Equilibration of Diluted Semen

The diluted semen is filled in 0.25ml poly vinyl chloride straws using an automatic filling and sealing machine (IS-4, IMV Technologies, France). Sealed straws are arranged on a rack and equilibrated for 4 h in cold handling unit (GO-38, IMV Technologies, France).

Freezing of Semen

Immediately after equilibration the rack containing filled semen straws are transferred to a bio-freezer (Digit cool-5300, IMV Technologies, France) where the temperature is brought down from 4 to -140° C in 7 minutes. The straws are then transferred to pre-cooled plastic goblets and plunged in to liquid nitrogen for storage. Cryopreserved semen straws are kept immersed in liquid nitrogen till use.

Thawing of Semen

Frozen semen straws are thawed at 37° C for 30 seconds for its use in artificial insemination programme.

Table 1: Effect of Extenders on Semen Characteristics of Diluted Semen (Mean ± SE)

Parameter	*Type of Diluent*	*Jersey Bulls*	*Buffalo Bulls*
Progressive Motility (per cent)	TEY	71.38±0.54 a	72.70±0.86 a
	SE	66.77±0.53 b	67.12±0.84 b
LiveSperm (per cent)	TEY	77.05±0.73 a	76.37±0.36 a
	SE	74.94±0.67 b	72.89±0.42 b
Head abnormalities (per cent)	TEY	3.70±0.20 NS	3.82±0.32 NS
	SE	3.83±0.20 NS	4.17±0.38 NS
Mid-pieceabnormalities (per cent)	TEY	3.72±0.19 NS	3.36±0.31 NS
	SE	4.05±0.22 NS	3.83±0.36 NS

Parameter	*Type of Diluent*	*Jersey Bulls*	*Buffalo Bulls*
Tail abnormalities (per cent)	TEY	5.66±0.19 NS	5.86±0.32 NS
	SE	6.33±0.26 NS	6.59±0.44 NS
Total sperm abnormalities (per cent)	TEY	13.16±0.36NS	13.05±0.70NS
	SE	14.22±0.51NS	14.60±0.80NS
Acrosome integrity (per cent)	TEY	84.16±0.60NS	85.52±0.52NS
	SE	83.83±0.62NS	84.44±0.52NS

Table 2: Semen Characteristics of Equilibrated Semen (Mean ± SE)

Effect of Extenders on Semen Characteristics of Equilibrated Semen (Mean ± SE)

Parameter	*Type of Diluent*	*Jersey Bulls*	*Buffalo Bulls*
Progressive Motility (per cent)	TEY	66.38±0.54 a	68.34±0.79 a
	SE	61.61±0.55 b	63.52±0.85 b
Live Sperm (per cent)	TEY	72.00±0.80 a	71.69±0.53 a
	SE	68.44±0.84 b	68.20±0.49 b
Head abnormalities (per cent)	TEY	4.00±0.26 NS	4.11±0.42 NS
	SE	4.10±0.25 NS	4.27±0.39 NS
Mid-piece abnormalities(per cent)	TEY	4.27±0.21 NS	4.03±0.33 NS
	SE	4.83±0.24 NS	4.79±0.41 NS
Tail abnormalities (per cent)	TEY	6.77±0.19a	6.63±0.23a
	SE	7.55±0.24 b	7.18±0.44 b
Total sperm abnormalities(per cent)	TEY	15.05±0.53NS	14.77±0.85NS
	SE	16.50±0.60NS	16.25±0.98NS
Acrosome integrity (per cent)	TEY	80.77±0.73NS	82.41±0.81NS
	SE	80.00±0.85NS	81.25±0.98NS

Table 3: Effect of extenders on semen characteristics of post thaw semen (Mean ± SE)

Parameter	*Type of Diluent*	*Jersey Bulls*	*Buffalo Bulls*
Progressive Motility (per cent)	TEY	43.05±0.82a	44.65±1.25a
	SE	38.33±0.80b	38.19±1.15b
Live Sperm (per cent)	TEY	51.16±1.00a	51.98±0.71a
	SE	47.50±1.06 b	47.26±1.05 b
Head abnormalities (per cent)	TEY	4.27±0.27 NS	4.33±0.58 NS
	SE	4.44±0.25 NS	4.66±0.48 NS

Parameter	*Type of Diluent*	*Jersey Bulls*	*Buffalo Bulls*
Mid-pieceabnormalities (per cent)	TEY	4.83±0.20 NS	4.69±0.45 NS
	SE	5.33±0.24 NS	5.33±0.45 NS
Tail abnormalities (per cent)	TEY	7.77±0.19 a	7.84±0.31 a
	SE	8.61±0.23 b	8.59±0.47 b
Total sperm abnormalities (per cent)	TEY	16.88±0.46 a	16.86±0.69 a
	SE	18.44±0.49 b	18.55±0.78 b
Acrosome integrity (per cent)	TEY	77.27±0.77 NS	78.35±1.21 NS
	SE	76.16±0.79 NS	77.53±1.21 NS

Recommendations

1. Efficacy of Soya Based Extender was Tested on Cryopreservation of Jersey Bull Semen

Soya based extender resulted in acceptable freezability of cattle semen in terms of motility, live sperm count, intact acrosomes and sperm abnormalities. Although there were differences in all the parameter compared between Tris egg yolk (TEY) and Soya based extender (SBE) yet, Tris egg yolk exhibited better results. But these differences are in acceptable limits in Soya based extender. Hence, proposed extender could be recommended for use in cryo-preservation of bovine semen.

2. Effect of Glutathione and Cysteine Hydrochloride as Semen Additive was Tested on Cryopreservation of Crossbred Bull Sperm using Soya Milk Extender

Cysteine, an amino acid precursor of glutathione is considered to be an intracellular antioxidant which protects the cells from reactive oxygen species (ROS) mediated damage under oxidative stress. Glutathione is a tripeptide and is present mainly in reduced form (GSH) and only a small amount is in oxidized form (GSSG). It stimulates the aerobic glycolysis and aerobic respiration of epididymal sperm in semen. Effect of soya based extender with glutathione (SBE-G) @ 5mM and soya based extender with the addition of cysteine hydrochloride (SBE-CY) @ 2mM was tested on cryopreservation of crossbred bull sperm. The results showed that post-thaw semen characteristics improved with addition of these additives as compared to non-added diluent. In-vitro fertilizing potential of spermatozoa also improved with the addition of these additives. Post-thaw acrosome integrity (per cent) and HOST positive (per cent) spermatozoa were highest in glutathione added diluent followed by cysteine added diluent. Glutathione @ 5 mM and cysteine hydrochloride @ 2 mM can be safely added as additives to semen in soya based extender to improve the post-thaw semen characteristics.

3. Effect of Enzymatic Anti-oxidants in Soya Milk Extender on Cryo-preservation of Buffalo Semen

Addition of superoxide dismutase (SOD) and catalase (CAT) in soya based extender (SBE) was tested on cryopreservation of buffalo semen. The results showed that addition of SOD and CAT to buffalo bull semen freezing extender causes significant increase of post thaw sperm motility, membrane integrity, and acrosome integrity. *In-vitro* fertilizing potential of spermatozoa showed significant improvement in TEST (with SOD and CAT) diluent when compared to CONTROL (without additive). SOD and CAT @ 100units/ml can be safely used as additives to buffalo semen in soya based extender to improve the post-thaw semen characteristics. SOD and CAT enzyme supplementation to extender can contribute greatly to the reduction of sperm membrane lipid peroxidation by ROS during buffalo semen cryopreservation and thus allow good semen quality after freezing. After addition of antioxidants to SBE its quality is improved.

Benefits/Impact

Semen extended with soya based extender (SE) can be an alternative to the conventional Tris extender as it is free from animal proteins.

Division of Veterinary Pathology

80. Molecular Characterization and Phylogenetic Analysis of different Avipox Viruses

Contact Scientist: **Dr. Nawab Nashiruddullah**

Avipox viruses were isolated from different bird species in Jammu region to find their phylogenetic relatedness.

Methodology

For diagnosis and confirmation of *Avipoxvirus*, the *P4b* gene encoding for 4b core protein of the virus was targeted. The primer sequences used in PCR was as previously described by Lee and Lee (1997). *Fowlpoxvirus, Pigeonpoxvirus* and *Turkeypoxvius* was sequenced and aligned by CLUSTAL W programme of Mega 5 with published sequences obtained from GenBank, and phylogenetic relationships were determined based on these alignments.

Phylogenetic analysis showed Fowlpox and Turkeypox virus isolates to cluster under recognized clade A1 along with other *Fowlpoxvirus* strains reported from India and other parts of world; however, Pigeonpox isolate clustered under established clade A2 where majority of *Pigeonpoxvirus* have been reported. The divergence and relatedness analysis showed Jammu Fowlpox and Pigeonpox virus isolates to be 90 per cent identical, while Fowlpox virus and Turkeypox virus showed 99 per cent identity. The sequences of Fowlpox, Pigeonpox and Turkeypox viruses have been published in GenBank with accession nos. MF496042, MF496043, MF496044 respectively.

Fowlpox Pigeonpox Turkeypox

Fowlpox, Pigeonpox and Turkeypox in Birds from Jammu.

Recommendations

More *Avipoxvirus* need to be isolated from birds so that the virus diversity is recognized within this region.

Benefits/Impact

- This is the first isolation and molecular characterization of *Avipoxvirus* in the state.
- Molecular characterization and phylogenetic analysis of viruses help in molecular epidemiology.

81. AgNOR as a Molecular Marker of Cell Proliferation in Various Tumors of Domestic Animals

Contact Scientist: **Dr. Shafiqur Rahman**

Methodology

AgNOR study was conducted on the specimen collected for histopathology. For each case, a tissue section of tumour was stained for AgNOR by using the standard method with some modifications.

Name of Reagents	*Amount of Reagents*	*Concentration*
Stock Solution A		
Gelatin powder	2 gm	2 per cent
Formic acid	1 ml	
Deionised water	100 ml	
Stock Solution B		
Silver nitrate	50 gm	50 per cent
Working Solution		
Solution A	1 part	
Solution B	2 parts	

Protocol

Formalin fixed 4-5 μ paraffin sections were stained with a solution made up of 50 per cent silver nitrate solution (prepared fresh) and 2 per cent gelatine stored at room temperature. Working solution was freshly prepared using two parts of silver nitrate solution and one part of gelatine solution. To carry out AgNOR staining, the cut sections were deparaffinised and rehydrated using distilled de-ionized water. Then place in plastic jar filled with the staining solution and incubated at 37°C in dark for 60 minutes and washed thoroughly with distilled deionised water and dehydrated. After drying, the slides were cleared in xylene and mounted in DPX mountant.

The AgNORs were visualized as brown black discrete dots of variable size within the nuclei. AgNOR in 100 cells was counted and size variation was also recorded. Areas with necrosis, pronounced inflammation, artificial damage and marked keratinisation were avoided. Black dots (both smaller and larger) corresponding to clusters was treated as one AgNOR method.

Table 1: Showing mAgNOR Count per Nuclei (Mean ± SD) and pAgNOR (per cent) in Bovine Neoplasms

Types of Tumour	*mAgNOR count(Mean ± SD)*	*pAgNOR count (per cent)*
Benign Tumours		
Fibroma	2.24 ±0.98	14
Cutaneous papillomatosis	1.94 ±0.84	10
Myxoma	1.72 ±0.67	8
Epulis	1.80 ±0.79	10
Ductal adenoma	1.61±0.88	8
Fibroma	2.24 ±0.98	14
Malignant Tumours		
Fibrosarcoma of skin	2.84±1.13	34
Cutaneous squamous cell carcinoma	2.72±1.05	30
Ocular Squamous Cell Carcinoma	2.83 ±0.87	32

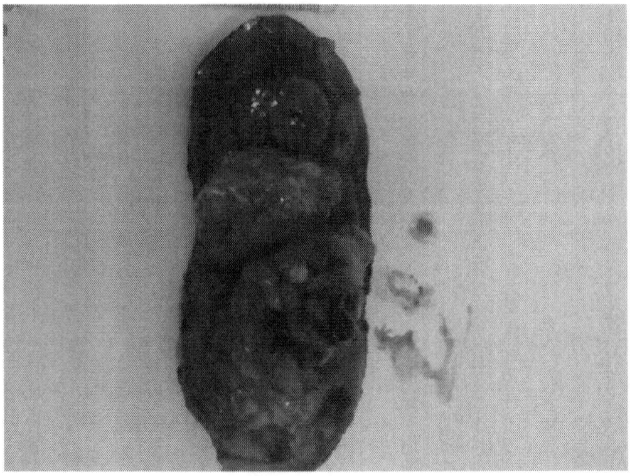

A Large, Irregular, Hard, Nodular, Poorly Demarcated Tumour Mass.

Recommendations

The mAgNOR count per nuclei and pAgNOR were both lower in benign tumour than malignant tumour in domestic animal species. Rapidity of cell proliferation- a characteristic of malignant tumour can be studied very well by this method.

Benefits/Impact

- This method will help to veterinarians for prompt diagnosis, prognosis and treatment of neoplasms.
- It will help to rule out the other conditions mimicking with tumours like chronic abscess, cyst, parasitic nodules *etc.*
- It was found to be a better indicator to assess rapidity of cell proliferation (a hall mark of malignant tumour) than Histopathology.

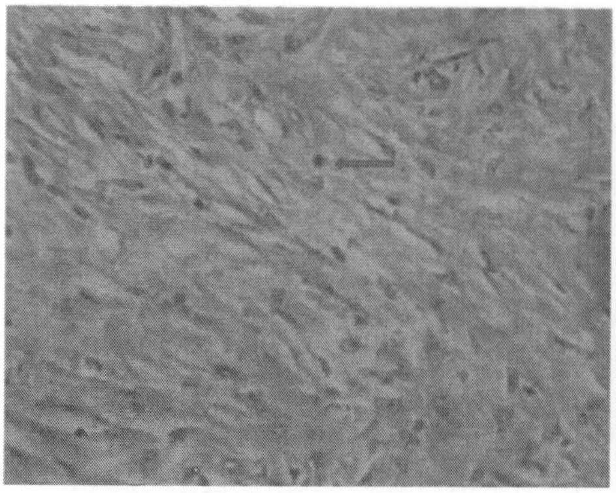

Photomicrograph Showing Pleomorphic Anaplastic Cells with Elongated Hyperchromic Nuclei that Vary in Size and Shape. Note the presence of mitotic figures (arrow). H and E X 400.

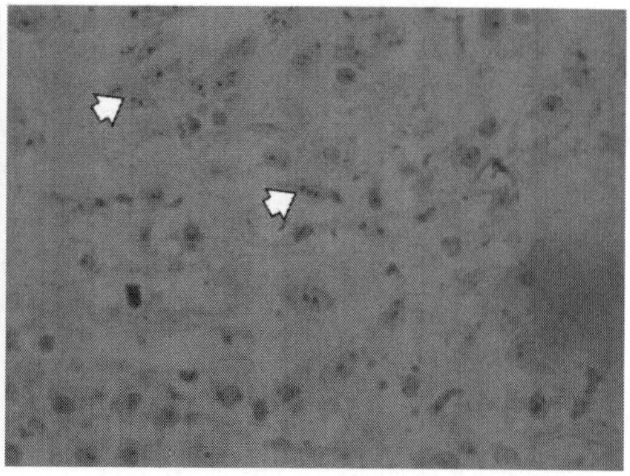

Fibrosarcoma Showing Numerous AgNOR Dots (Arrows) Dispersed in the Nuclei. AgNOR X 400.

82. Etio-patho Morphological Studies on Gastrointestinal Tract of Broiler Chickens in Jammu

Contact Scientist: **Dr. Shilpa Sood**

Methodology

A study was carried out determine etio-pathomorphology of gastro-intestinal tract of broiler chickens. The study period was divided into four seasons from July 2016 to June 2017 as per Meteorological Department of India, Pune, *viz.*, monsoon (July to September), post monsoon (October to November), winter (December to February) and summer (March to June).

A total of 200 different poultry flocks in and around Jammu where disease occurrence was reported were included in the study. Age wise and season wise disease occurrence and mortality pattern at the field level was determined.

Occurrence of disease percentage was calculated as:

$$\text{Occurrence of disease} = \frac{\text{Total no. of morbidity and Total no. of mortality}}{\text{Total no. of birds in the flock}} \times 100$$

A total of 632 birds were necropsied. Systemic examination was carried out for the lesions in the GI tract including esophagus, proventriculus, gizzard, liver, intestine and pancreas.

Occurrence of lesions was calculated as follows:

$$\text{Occurrence of lesions in organ} = \frac{\text{Total no. of carcasses showing lesion in that organ}}{\text{Total no. of birds necropsied}} \times 100$$

Representative samples from the necropsied birds were collected for bacterial or parasitic examination. Tissue samples of various organs were collected in 10 per cent neutral buffered formalin and processed for routine histopathological examination. Liver and heart were collected from dead birds aseptically into sterilized petri-plates for bacteriological examination. After complete biochemical and microbiological characterization, all the isolates of *E. coli* and *Salmonella* were sent for serotyping to the National *Salmonella* and *Escherichia* centre, CRI, Kasauli (HP), India.

Table 1: Percentage of Occurrence and Mortality Pattern of Various Diseases in Broilers at different Farms

Disease	*Total No. of Flocks*	*Total Number of Birds*	*Total Number of Morbidity (Approx.)*	*Total Number of Dead Birds*	*Occurrence (Per cent)*	*Mortality (per cent)*
E. coli	84	180753	126000	1500	22.57	1.19
Omphalitis	17	53900	36685	238	6.53	0.64
Salmonella	24	50600	29250	153	5.20	0.52
IBD	22	44800	38000	284	6.77	0.74
Coccidiosis	06	21600	12380	30	2.19	0.24
Haemorrhages in intestines/ proventriculus (ND/AI suspected)	03	11500	8980	62	1.60	0.69
Caseous nodules in lungs (Brooder Pneumonia suspected)	03	11000	4200	12	0.74	0.28
Nephro-toxicity	06	13000	7830	33	1.39	0.42
Gout	05	24500	8100	07	1.43	0.08
Ascites	07	17200	6400	09	1.13	0.14
Enteritis	16	35600	23485	113	4.17	0.48
Fatty liver	04	42500	5800	03	1.02	0.05
Hepatitis	02	4700	2765	12	0.49	0.43
Internal haemorrhage	03	36400	6900	05	1.22	0.07

Occurrence of colibacillosis (24.16 per cent) was maximum followed by that of infectious bursal disease (IBD) (7.13 per cent), omphalitis (6.79 per cent), salmonellosis (5.36 per cent), non specific enteritis (4.29 per cent), coccidiosis (2.22 per cent), haemorrhages in proventriculus (1.67 per cent), gout (1.44 per cent), nephro-hepatotoxicity (1.42 per cent), ascites (1.14), fatty liver syndrome (1.03 per cent), caseous nodules in lungs/suspected brooder pneumonia (0.75 per cent) and non specific liver ailments (0.50 per cent).

Recommendations

- Various diseases are prevalent among broiler population of Jammu that affect the digestive tract were identified for which effective control measures must be put in place to check the spread of diseases from one farm to another.

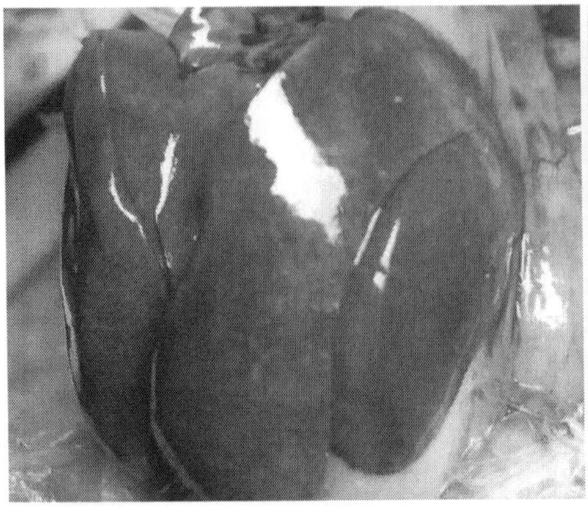

Greenish-Bronze Discoloration of Liver

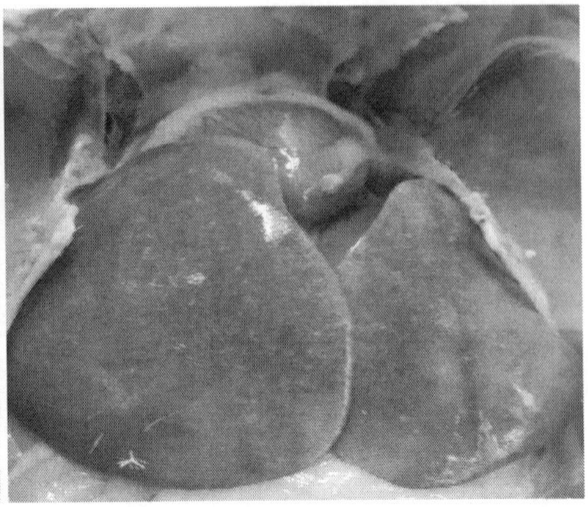

Presence of Necrotic Foci in Liver

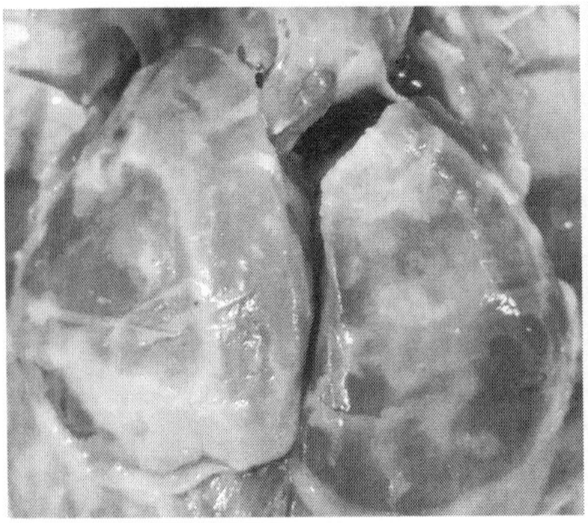

Fibrinous Perihepatitis

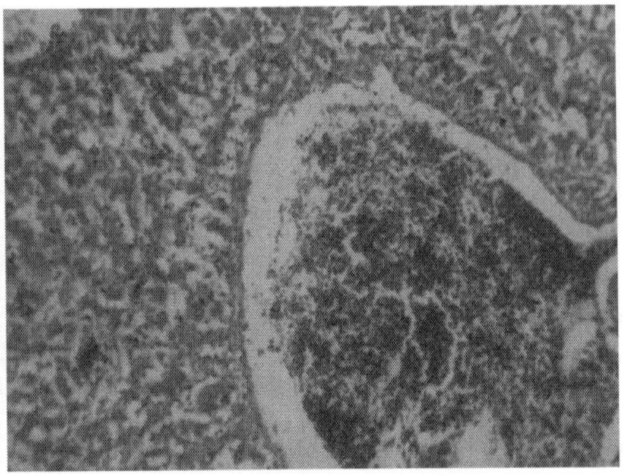

Severe Congestion and Degeneration in Liver (400X H and E)

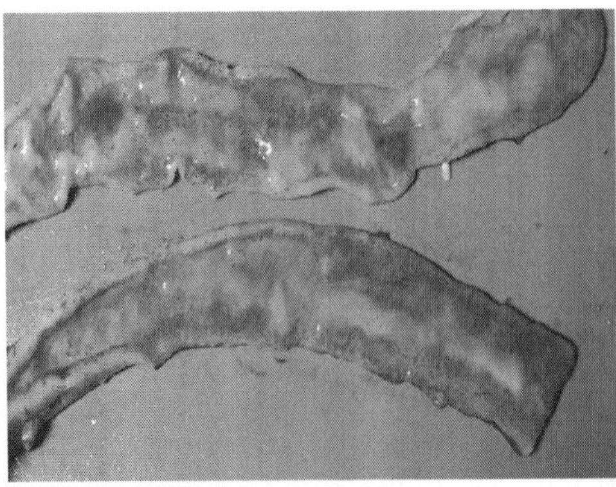

Severe Enteritis

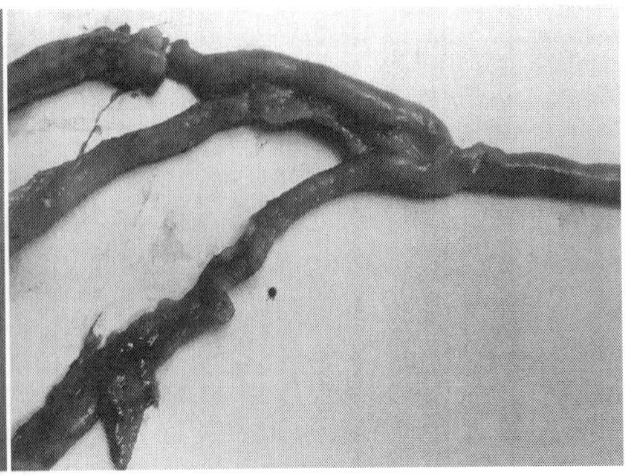

Haemorrhages in Caeca

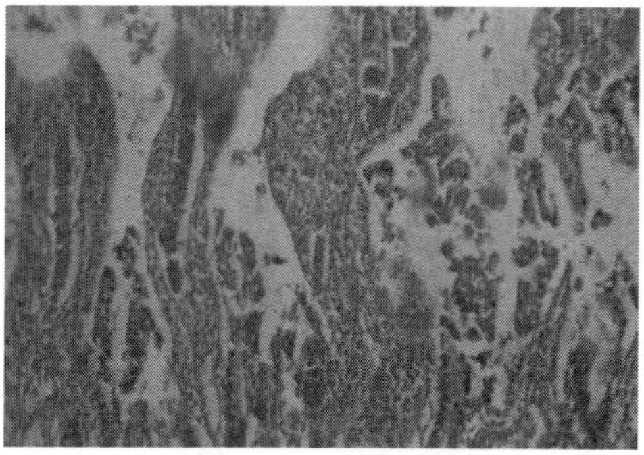

Haemorrhagic Enteritis

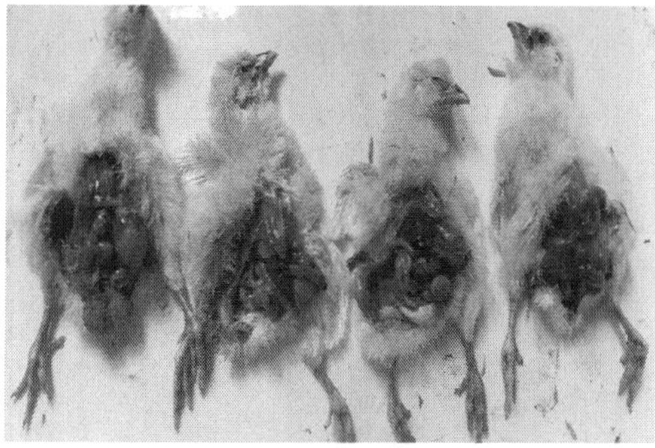

Omphalitis

- Farmers need to be educated about the mode of transmission of these diseases.
- Awareness among the farmers must be created regarding the various vaccination programs to be undertaken for prevention of the occurrence and spread of these diseases.
- After accurate diagnosis, relevant and appropriate treatment measures can be instituted

Benefits/Impact of Technology

- Diagnosis of diseases using pathological and microbiological examination help in identification of specific lesions and diseases affecting gastro-intestinal tract of broiler chickens.
- The study was helpful to identify different diseases prevalent in different seasons affecting various target age groups.
- The data collected from the present study will help in formulation of measures for effective treatment, prevention and control of diseases particularly those affecting digestive tract of broilers.

Division of Veterinary Public Health and Epidemiology

83. Scientific Interventions for Control of Zoonotic Parasites of Livestock and Livestock Rearers

***Contact Scientist:* Dr. Mohd. Rashid**

Introduction

The livestock farming play an important role in state economy and is an important subsector of agriculture. The total livestock population of state was 96.39 lakhs as per 19th. Livestock Survey of India. It plays a vital role in providing nutritive food, rich in animal protein to the general public and in supplementing family incomes and generating gainful employment particularly to the nomads and rural youths. These people are rearing vast herds/flocks which form their chief dependence and trafficking into remote areas under great hardship along with their off-springs.

Majority of the herds/flocks migrate from plain to high altitude alpine pastures but very few remain sedentary. During migration these animals are attacked by wide array of ectoparaites and endoparasites. Thus scientific interventions for control of zoonotic parasites of sheep and goat should be an important part of flock health managements that should be implemented at the door steps of rearing community by active participation.

Parasites are one of the main factors that affect the productivity of livestock especially in the wet and highland areas. The economic losses due to parasitic diseases to meat industry in J&K state are not estimated but they are in millions annually. Besides this the zoonotic diseases are also making vagaries to the rearing community due to close association and lack of epidemiology of diseases. Thus these interventional strategies are intended to improve the socioeconomy of farming community by improving productivity through the control of zoonotic parasitism both in animal and human beings.

Methodologies

The scientific interventional strategies adopted for effective control of zoonotic parasitic diseases are:

1. Educational awareness in respect of parasitic and other common zoonotic diseases by way of organising camps, distributing reading materials (pamphlets, leaflets, bulletins *etc.* in English and local language Urdu/ Hindi) to sensitise the prevention and control of zoonoses among livestock and livestock rearers.
2. Training of the livestock farmers at the door step with the objectives:
   a. To introduce participants the principles and concepts of adult learning and its importance in control of diseases.
   b. To pass the knowledge about the transmission of diseases and their control measures.
   c. To develop participant's capacity to improve livestock and human health
   d. To develop participant's capacity to formulate achievement based objectives
   e. To equip participants with knowledge of effective training methods for others.
3. Formulation of action plans keeping in mind the climatic conditions of different regions of state. As in Jammu and Kashmir livestock is reared by nomads and are kept on pasture throughout the year and climatic conditions are favourable for the development and survival of infective stages or vectors. Helminthic parasites are therefore a major cause of economic loss directly or indirectly by lowering the immune status of animals leading to vagaries of other diseases. Major factors that contribute to the multiplication and sustainability of parasites are climate, ecology and systems of animal production. As a result, the distribution and intensity of infection of helminthic parasites vary from area to area depending on these factors. So it is not possible to recommend one deworming program suitable for all regions of Jammu and Kashmir. Local knowledge is essential to develop such programs. So treatment (deworming) of livestock with appropriate drugs will be suggested in different quarters.

Table (data)

Table 1: Quarter-wise Detail of Liaison Campaigns and Awareness-cum-Treatment Camps

	Q. 1	Q. 2	Q. 3	Q. 4	Q. 5	Q. 6	Q.7	Q.8	Total
No. of Liaison campaigns organised	7	3	5	8	2	1	5	2	**33**
No. of people participated in Campaigns	120	28	50	100	60	32	200	-	**590**
No. of Resource persons involved	5	7	5	6	5	3	4	2	**37**
No. of treatment camps conducted	9	4	7	7	6	2	5	2	**42**
No. of people participated in awareness cum treatment camps	295	150	60	300	130	50	270	125	**1380**
No. of Resource persons involved	5	6	7	10	5	3	6	3	**45**

Name of Village	Liaison Campaigns	Awareness cum Treatment Camps	No. of Participants
Fazalabad	1	2	70
Muchakki Draba	1	2	80
Pathian Draba	1	1	30
Daddanar Surankote	1	1	10
Surankote Samote	1	1	10
Chingus Rajouri	1	2	20
Kura Kalakote	1	2	60
Barmandal Kalakote	1	1	20
Kharakpanjah Kalakote	2	3	90
Kurlian Kalakote	1	1	40
Katwari Kalakote	2	2	155
Hadayatpur Narian	1	2	50
Rajalmore Nowshara	1	2	45
Chattyari Nowshara	1	1	20
Dogiani Nowshara	1	1	25
Sangpur Nowshara	1	1	35

Name of Village	Liaison Campaigns	Awareness cum Treatment Camps	No. of Participants
Sunderbani	1	1	10
Tanddi Surankote	2	2	25
Rajalkote Nowshera	1	1	40
Rajalmore Nowshera	1	2	70
Gurn Nowshera	1	1	15
Sarrah Kalakote	1	1	20
Kura Gujran Kalakote	2	2	70
Chainpur Kalakote	1	1	30
Dalhote Kalakote	2	2	100
Lathung Surankote	1	2	70
Keri Barmandal	1	1	80
Broh Barmandal	1	1	60
Total	**33**	**42**	**1380**

Recommendations

1. Continuous surveillance and monitoring of parasitic diseases among livestock and livestock farmers to eliminate the zoonotic parasitic diseases burden.
2. Frequent conducting of awareness and training programme by intersectoral coordination at the door steps of farming community.

Benefits/Impact

Benefits of Awareness Camps

In the awareness camps both the male and female are invited to participate. Well prepared lectures/talks were delivered in local language on the damages caused by external and internal parasites, such as damage to host skin, allergic reaction, ulceration, anaemia, annoyance, disturbed feeding, transmission of disease causing organisms of other biological origin.

They were also educated about the transmission of parasites to other livestock and human population, maintenance, breeding, survivability in the environment. They rear dogs along their flocks for watch and ward purpose and horses for transportation of their livelihood materials, which also share these zoonotic diseases.

They were finely taught how to control these parasites among livestock, human beings and in the environment, so that they could not act as carrier or reservoir of other diseases besides causing direct losses to livestock industry.

Benefits of Trainings

a. Participants learned the principles and concepts of adult learning and realized the gaining of knowledge in control of diseases.
b. Participants gained the knowledge about the transmission of diseases and their control measures.
c. Participants developed capacity to improve livestock and human health
d. Participants developed capacity to formulate and implement action plan
e. Participants developed equipped with knowledge of effective training methods for others.

Benefits of Distribution of Pamphlets and Bulletins

Pamphlets were written in local Urdu language with all the details that we teach them with the help of poster and oral presentation with the objective to further disseminate the knowledge about the harmful effects of parasites and their prevention/control strategies through the help of educated youths or school going children. In this way they also get literature for further dissemination of knowledge to others.

Benefits of Distribution of Drugs/Anthelmintics

These were used for treating the infested animals to overcome damages and losses caused by the parasites as enlisted earlier. Effective treatment of positive animals reduce the sheding of eggs leading to decrease of incidence of parasites from pastures *etc.* Moreover no one comes to listen speeches without incentives. Drugs acts as bridges between us and farmers, by virtue of treatment they follow the techniques of application and note down withdrawal time. As most of the acaricides are carcinogens if consumed through milk or meat or through their products. So once they are awaked about these life threatening problems they take care of withdrawal period and non judicial use of acarisides.

Impact

Detail of impact with respect to some of the questionnaires ranged from 50 per cent to 95 per cent as predicted below in table but the mean impact was 79.25, which is a very good response.

Table 2: Impact of Study in Terms of Increased Knowledge in Percentage (per cent)

Sl. No.	*Questionnaires'*	*At the Time of Start of Study (per cent)*	*At the End of Study (per cent)*	*Impact (per cent)*
1	Do the diseases of animals will spread among the human beings?	5	95	94.73
2	Do the eggs of parasites are in dung of animals?	0	95	95.00
3	Do the parasites can spread through dung?	0	95	95.00
4	Do the wild animals and dogs help in perpetuation of parasitic diseases among the human beings?	0	90	90.00
5	Do we go for identification of external and internal parasites of animals and man? If yes why?	0	85	85.00
6	Do the medicines/antelmintics used for treating external and internal parasites are excreted through milk?	0	80	80.00
7	Do we consume milk and meat after immediate treatment of animals?	0	80	80.00
8	Do we use under doses of anthelmintics on animals?	0	90	80.00
9	What are the harmful effects of under doses of drugs/anthelmintics	10	98	89.79
10	Encircle the diseases which spread from animals to man from the given lists	5	90	94.44

Sl. No.	*Questionnaires'*	*At the Time of Start of Study (per cent)*	*At the End of Study (per cent)*	*Impact (per cent)*
11	At what time we should use anti- parasitic drugs on animals before or after drinking of water	7	97	92.78
12	Do we use anti-parasitic drugs on animals inside their sheds? If not why?	0	92	92.00
13	What we should do for prevention and control of parasitic diseases?	0	50	50.00
	Overall impact of study			**79.52**

84. Public Health Significance of *Echino coccosis* in Jammu

Contact Scientist: **Dr. M.A. Malik**

Methodology

In present study, screening of dogs (definitive host), sheep and goats (intermediate hosts) and humans (accidental host) for the presence of Echinococcus/metacestode infection was done. Faecal samples of 105 dogs (50 pet and 55 stray) were collected during study. Besides this necropsy examination of 20 stray dogs died accidently during the study period was carried out to find the adult parasites/worms in the intestines of dogs in and around the vicinity of suspected stray dogs population. For prevalence of Echinococcus cysts in sheep and goats (intermediate hosts), physical examination of visceral organs of about 425 sheep (n=230) and goats (n=195) slaughtered in various slaughter houses of Jammu region was done for presence of echinococcus/hydatid cysts during postmortem inspection during the period of study. For Studies on sero-prevalence of hydatidosis in human beings, 92 samples (from patients with liver disorders, dog handlers, veterinarians, dairy farmers, *etc.*) were collected and subjected to IgG-ELISA for serum antibody detection (Shambesh *et al.*, , 1997) of hydatidosis.

Results

- In total 105 dog faecal samples were examined by direct and concentration methods and 5.71 per cent (n=6) were found positive for Taenid eggs.
- The prevalence of Taenid eggs in stray dogs was 16 per cent and zero per cent in pet dogs.
- Arecoline purge to six dogs positive for Taenid eggs @ 4mg per kg body weight revealed one dog positive for *Echinococcus granulosus.* The other parasitic eggs/segments observed were of *Ascarid eggs* (40/105), *Strongyle* eggs (5/105) and *Dipylidium caninum* (4/105).
- Necropsy of 20 stray dogs died in road accidents was performed during study period. Out of those twenty dogs 11 were male and 9 were female. During necropsy detailed inspection of intestinal content was conducted to find the worms. Out of 20 dogs, none of the dogs was found positive for *Echinococcus granulosus* showing zero per cent prevalence. However most of the dogs were found positive for other parasites of zoonotic importance *viz.* eight dogs were positive for *Ancylostoma caninum* (40 per cent), three for *Toxocara canis* (15 per cent), two for *Taenia hydatigena* (10 per cent), and four were positive for *Dipylidium caninum* (20 per cent).

- In slaughter house studies on 425 animals, 6.82 per cent (29/425) animals were positive for hydatid cysts. The sheep showed slightly higher (7.83 per cent) prevalence as compared to goats (5.64 per cent).
- In seroprevalence studies on human hydatidosis using IgG-ELISA kit, 14.13 per cent (13/92) samples were positivite. The seroprevalence in different risk groups was observed to be 33.33 per cent in dog owners/handlers, 25 per cent in vegetable vendors, 21.42 per cent in dairy farmers/farm labour, 11.53 per cent in patient with liver disorders, 11.11 per cent in others and zero per cent in veterinary doctors.

Collecion of Intestine

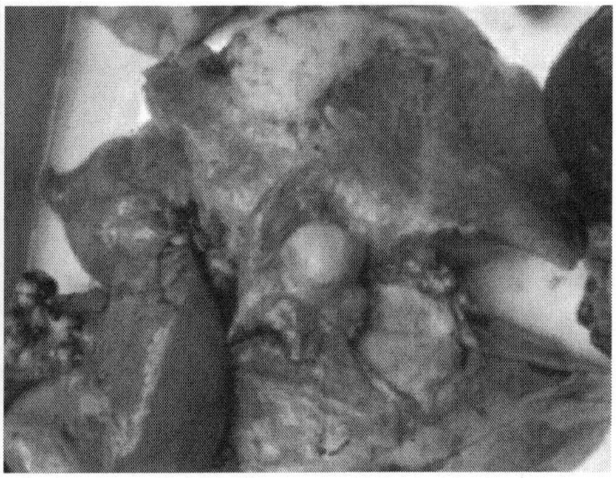

Multiple Hydatid Cysts in Lungs of Sheep

Impact/Recommendations

The study concludes that stray dogs of Jammu have high potential to transmit zoonotic diseases of parasitic origin to humans. High prevalence of fertile hydatid cysts in sheep and goats suggests alarming indications for proper disposal of viscera of infected small ruminants as stray dogs have easy access to condemned meat/offal. High seroprevalence using Ig-G ELISA in human individuals of different risk groups suggests that suitable control measures are to be adopted at earliest like, control of population of stray dogs, proper disposal of condemned meat/offals, awareness campaign among people about the disease dynamics and hygienic measures to be adopted, routine deworming of pet dogs *etc.*

85. Enterotoxigenic Profile of *Bacillus cereus* in Meat and Meat Products

Contact Scientist: **Dr. M.A. Malik**

Bacillus cereus is a Gram positive, endospore forming bacteria that is ubiquitous in nature and is frequently associated with diarrheal and emetic type of food-borne illness. The diarrheal type is associated with meat, milk, soups, sauces and vegetables. The objective of the present study was to assess the incidence of *Bacillus cereus* in meat and meat products along with antibiogram pattern and molecular characterization with respect to enterotoxigenic profile.

Methodology

In the present study, a total of 150 samples of meat and meat products (50 raw mutton, 50 raw chicken and chicken biryani) from places in and around Jammu were analysed for the presence of *B. cereus* targeting *gyrB* gene. *B. cereus* isolates were characterized for the presence and distribution of enterotoxogenic gene complexes *viz.*, the hemolysinBL (*hblCAD*) complex and the non-hemolytic (*nheABC*) complex.

Results

- Out of 150 samples, 44 (29.33 per cent) samples yielded *B. cereus*.
- The highest incidence was observed in chicken biryani (36 per cent) followed by raw mutton (28 per cent) and raw chicken (24 per cent). However, cooked meat revealed a higher incidence of (36 per cent) than raw meat (26 per cent).
- All the isolates carried at least one enterotoxin gene. Further it was found that at least one gene of NHE complex was present in 39 (88.6 per cent) isolates. However at least one gene of HBL complex was present in 28 (63.63 per cent) isolates.

Impact/Recommendations

- Occurrence of *B. Cereus* in foods especially in meat and meat products was frequent with overall incidence of 29.33 per cent.
- Incidence was more in cooked meat 36 per cent than raw meat 26 per cent which is of public health concern.
- Molecular confirmation by detecting *gyrB* gene is helpful in differentiating *B. cereus* from rest of the members of the *B. cereus* group.
- All the isolates carried at least one toxin gene which directly indicates the pathogenicity potential of *B. cereus*.

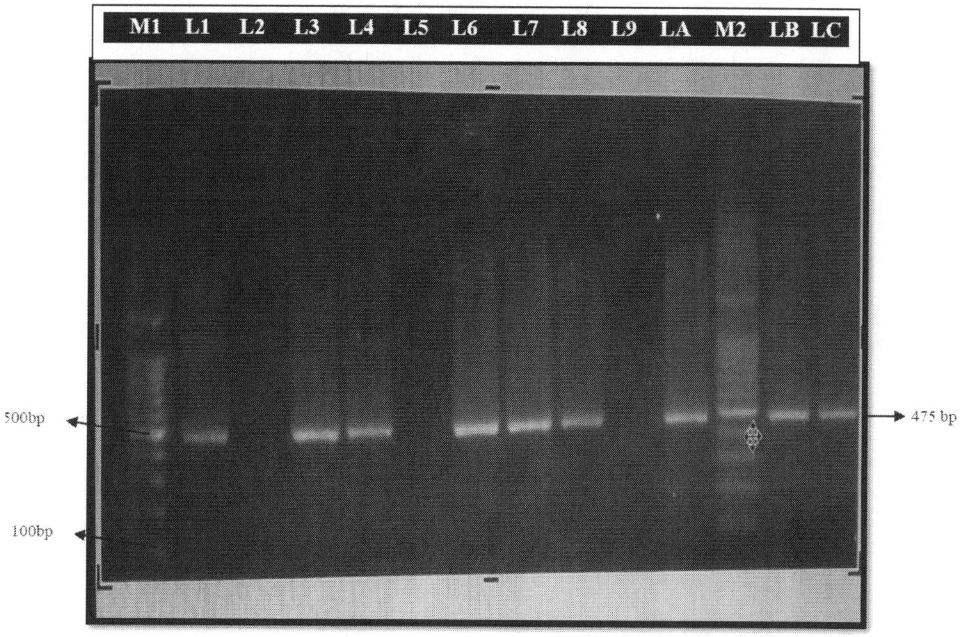

Plate-10: Agarose gel electrophoresis depicting amplified PCR product of *gyrB* gene from genomic DNA of *B. cereus* for confirmation

M1and **M2**-100 bp DNA ladder

L2, **L5** and **L9** – Negative controls

LA-Positive control

L1, **L3**, **L4**, **L6**, **L7**, **L8**, **LB** and **LC**-Samples positive for *gyrB* gene

- *B. cereus* were resistant to various antimicrobials like penicillin G, carbenicillin, ampicillin and amoxicillin which further signifies its public health importance owing to its enterotoxogenic potential.

The effective use of quality control and safety measures during production/ manufacture besides creating awareness among food handlers and consumers becomes necessary to ward off the menace of *B. cereus* and other food poisoning organisms.

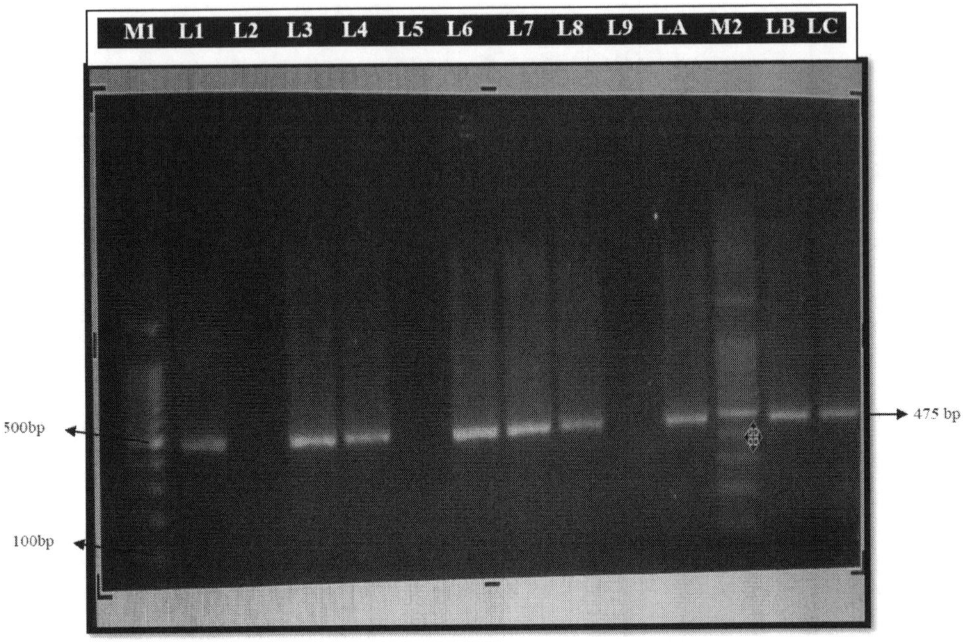

Plate-10: Agarose gel electrophoresis depicting amplified PCR product of *gyrB* gene from genomic DNA of *B. cereus* for confirmation

M1and **M2**-100 bp DNA ladder

L2, **L5** and **L9** – Negative controls

LA-Positive control

L1, **L3**, **L4**, **L6**, **L7**, **L8**, **LB** and **LC**-Samples positive for *gyrB* gene

Division of LPM

86. Heat Stress Amelioration in Dairy Cattle by Showering

Contact Scientist: Dr. Asma Khan

Methodology

Fifteen crossbred dairy cows were randomly selected and allotted to three treatment groups of five each, *viz.*, T0 group (control group), T1 group in which showering was done once in a day in between 1100 hrs to 1300 hrs and T2 group in which showering was done twice in a day in between 1100 hrs to 1300 hrs and 1600 hrs to 1800 hrs. The whole experiment was conducted under completely randomized design (CRD). All the animals were kept under similar feeding and managemental conditions by individually housing them in well-ventilated cement floor shed. Animals were allowed for four hour grazing daily in the early morning. Feeding was done as per standard, while water was provided ad libitum.

During the experimental period microclimatic changes, THI, physiological parameters and blood biochemical parameters were measured. Analysis of variance was carried out to differentiate responses between treatments.

Table (Data)

Parameters	*T0*	*T1*	*T2*	*F Value*
Rectal temperature (afternoon)	39.91a ±0.36	39.10b ±0.45	39.17b ±0.46	131.19**
Respiration rate (afternoon)	62.02a ±3.69	50.88b ±3.67	52.87b ±3.75	58.17*
Pulse rate (afternoon)	90.20a±5.18	77.10b±4.68	78.58b±4.28	32.13**
Haemoglobin (g per cent)	8.60a±0.13	9.23b±0.13	9.31b±0.15	5.17*
Blood glucose (mg/dl)	64.96a±1.06	66.10b±0.95	68.54b±0.80	18.8*
Packed cell volume (per cent)	30.02a±3.21	32.96b±2.41	33.45b±2.36	6.27*
Total protein (g/dl)	7.32a±0.22	7.94b±0.17	7.98b±0.19	8.47*
Alkaline phosphatase (IU/litre)	30.31a±3.32	18.34b±3.42	15.54b±2.32	14.22*
Alanine amino-transferase (IU/litre)	96.24a ±6.24	76.31b ±7.52	64.40b±8.86	9.28*
Aspartate amino-transferase (IU/L)	101.14a ±5.10	82.10b±4.16	78.21b±3.72	7.22*

Recommendation

After showering, the animal physiological parameters were within the normal range for at least one hour, therefore making animal less susceptible to heat stress.

Showering as a simple managemental practice is advisable to farmers during summer season in sub-temperate regions.

Benefit/Impact

Showering to animals during extreme summer could ameliorate heat stress associated production loss and improve animal welfare.

87. Pine Leaves as Bedding Material Improves Economics and Performance of Broiler Chicks

Contact Scientist: **Dr. B. Brahma**

Methodology

Two experiments, one each during winter and summer season were conducted involving VENCOB broiler chicks (N=120). The chicks in each experiment were randomly distributed into 4 treatment groups. Each treatment group had 3 replicates of 10 nos. of chicks with average group weight. All chicks were reared up to the age of 42 days under identical conditions of housing and management except the variation in bedding materials. The paddy husk (PH), paddy straw (PS), pine leaves (PL) and combination of paddy straw and pine leaves were used as a different bedding materials in the winter and summer experiments. The evaluation was made on the basis of the growth performance, carcass characteristics, litter quality, microbial and parasitic growth of litter, manurial value and welfare of the birds.

Economics of Production

Particulars	*T0*	*T1*	*T2*	*T3*
Av. No of birds	30	30	30	30
Cost of chicks@ Rs 23/chick(Rs)	600	600	600	600
Mortality	0	0	0	0
Cost of litter(Rs)	105	42	0	24
Cost of feed @ Rs 30/kg (Rs)	3528	3240	3240	3222
Total input cost	4233	3882	3840	3846
Weight of birds (kg)	60.30	59.40	60.00	57.90
Income from birds @ Rs 100/kg	6030	5940	6000	5790
Total return over litter cost (Rs)	1797	2058	2160	1944
Total difference over control (Rs)	-	261	363	147
Difference over control/bird (Rs)	-	8.7	12.1	4.9

Overall Mean ± S.E of growth performance of chicks in two experiments.

Parameters	Treatment			
	T0	T1	T2	T3
Experiment I				
Av. body weight, g	2502 ± 42.06	2418±56.81	2433.60±36.43	2377.60±36.43
Body weight gain, g	2417.80±13.17	2333.96±14.52	2337.23±13.27	2295.13±14.85
Feed intake, g	4571.30 ± 45.61	4272.82±44.70	4397.63±37.23	4286.53±37.23
FCR	1.89±0.03	1.83±0.02	1.88 ± 0.04	1.86±0.02
PER	0.35±0.03	0.36±0.03	0.34 ± 0.01	0.34±0.04
EER	5.31±0.40	5.50±0.34	5.32 ± 0.42	5.11±0.32
Mortality, per cent	-	3.33±0.08	-	-
Experiment II				
Av. body weight, g	2018.83±31.11	1983.80±33.27	2007.36±35.73	1938.43±36.35
Body weight gain, g	1930.47±7.83	1895.33±5.34	1917.43±9.43	1848.93±4.38
Feed intake, g	3623.47±28.15	3604.33±32.66	3607.65±33.48	3588.63±26.86
FCR	1.88±0.02	1.90±0.04	1.88±0.02	1.94±0.03
PER	0.34±0.02	0.34±0.03	0.35±0.01	0.35±0.01
EER	5.18±0.26	5.15±0.43	5.30±0.35	5.35±0.27
Mortality, per cent	-	-	-	-

Recommendation

Performance of broiler chicks reared on paddy straw and pine leaves litter material gained body weight and FCR comparable with paddy husk in both the season. Manure value, parasitic and bacterial count was comparable in all the groups. No case of leg abnormalities or abnormal behavior was observed in birds confirming the comfort and welfare of the birds and pine leaves alone has a potential similar to an ideal litter material and could be used as an alternate litter material to promote the economical intensive poultry production system.

Benefit/Impact

Pine leaves have a potential as ideal litter material and could improve the economics of broiler rearing.

88. Enhancement of Growth Performance and Meat Quality of *Cyprinus carpio* by Chicken Manure

Contact Scientist: **Dr. Sahar Masood**

Methodology

The experiment was conducted in plastic tubs having the capacity of 100 litre. Live specimen of *C. carpio*, obtained from Government fish hatchery,Jammu were acclimatized for a month before experiment. During acclimatization, fish were fed with conventional feed (rice bran and mustard oil cake in 1:1 ratio) at the rate of 5 per cent body weight. The experimental tubs were cleaned, fresh soil was spread to 5cm height and treated with lime @ 250 kg/ha. After three days of liming, tubs were filled with water at 40 cm depth. Experimental water in tubs were treated with low(PT1) @8000 kg/ha, medium(PT2) @10,000 kg/ha and high(PT3) @ 12,000 kg/ha doses of chicken manure and also with low, medium and high dose of inorganic fertilizer(urea + SSP) @ 104 kg/ha + 155 kg/ha (IT1), @218 kg/ha + 310 kg/ha (IT2) and 322 kg/ha +470 kg/ha (IT3) respectively, along with control maintained on conventional feed without nutrient input. All the treatments were in duplicates. For water quality, temperature and pH (daily), DO and free CO2 (biweekly) and total solids, nitrate and phosphate (weekly) were determined as per APHA (1992). Sampling and processing of fish meat (AOAC, 1995) for proximate analysis and recording of length and weight to know SGR, condition factor was done initially and after 60 days at the end of the experiment. Fish were observed for mortality throughout the experiment to calculate the survival rate. The data were tested by using two way ANOVA. Significance was tested at 5 per cent level.

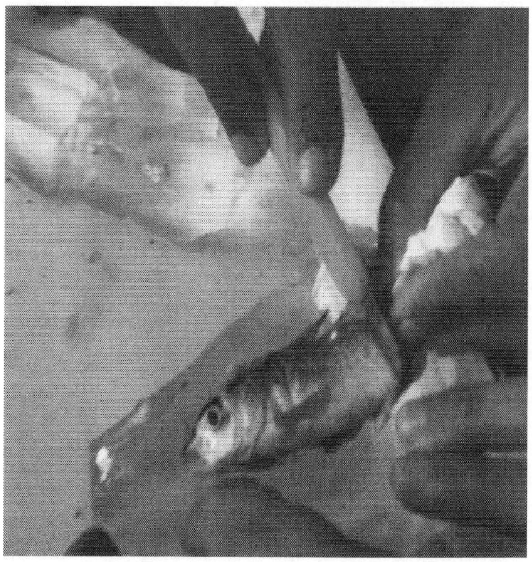

Table 1: Effect of Low, Medium and High Doses of Chicken Manure and Inorganic Fertilizer (SSP and urea) on Growth Parameters of *Cyprinus carpio* during 60 Days of Experiment (Mean ± S.E)

Parameters	*Treatments*						
	C	*PT_1*	*PT_2*	*PT_3*	*IT_1*	*IT_2*	*IT_3*
Initial weight (g)	$25.27^{a} \pm 0.50$	$27.25^{a} \pm 0.80$	$30.00^{a} \pm 0.50$	$30.17^{a} \pm 0.49$	$31.30^{a} \pm 0.15$	$25.91^{a} \pm 0.18$	$30.00^{a} \pm 0.35$
Final weight (g)	$25.75^{a} \pm 0.09$	$29.59^{b} \pm 0.75$	$35.10^{c} \pm 0.30$	$31.18^{b} \pm 2.44$	$32.31^{bc} \pm 0.88$	$28.11^{b}. \pm 0.48$	$32.25^{bc} \pm 1.0$
Initial Length (mm)	$87.6^{a} \pm 0.11$	$84.50^{a} \pm 0.02$	$87.70^{a} \pm 0.02$	$83.10^{a} \pm 0.14$	$85.40^{a} \pm 0.08$	$82.10^{a} \pm 0.20$	$82.6^{a} \pm 0.28$
Final length (mm)	$89.20^{a} \pm 0.17$	$86.00^{a} \pm 0.49$	$88.4^{a} \pm 0.48$	$84.30^{a} \pm 1.28$	$86.8^{a} \pm 0.15$	$83.00^{a} \pm 0.14$	$84.2^{a} \pm 0.15$
Specific growth rate (per cent)	$0.058^{a} \pm 0.11$	$0.099^{a} \pm 0.23$	$0.231^{c} \pm 0.53$	$0.074^{a} \pm 0.44$	$0.063^{a} \pm 0.21$	$0.182^{b} \pm 0.34$	$0.154^{b} \pm 1.00$
Initial Condition factor (K)	$4.15^{a} \pm 14.92$	$4.44^{a} \pm 22.0$	$4.28^{a} \pm 8.15$	$4.19^{a} \pm 8.03$	$4.51^{a} \pm 11.83$	$4.37^{a} \pm 15.93$	$4.21^{a} \pm 21.22$
Final Condition factor (K)	$4.14^{a} \pm 33.0$	$4.36^{a} \pm 21.34$	$4.86^{b} \pm 11.78$	$4.20^{a} \pm 13.75$	$4.44^{a} \pm 15.30$	$4.27^{a} \pm 25.00$	$4.21^{a} \pm 27.11$
Survival rate (per cent)	93.33	100.00	93.33	100.00	100.00	93.33	93.33

Mean bearing similar superscript with small alphabet in the row do not differ significantly with each other (Turkey HSD,$P<0.05$).

Table 2: Effect of Low, Medium and High Doses of Chicken Manure and Inorganic Fertilizer (SSP and urea) on Proximate Composition of *Cyprinus carpio* after 60 Days of Experiment (Mean ± S.E)

Parameters	*Treatments*						
	Control	*PT1*	*PT2*	*PT3*	*IT1*	*IT2*	*IT3*
Moisture (per cent)	$81.08^{a} \pm 0.15$	$81.01^{a} \pm 0.41$	$80.24^{a} \pm 0.24$	$80.91^{a} \pm 0.23$	$81.43^{a} \pm 0.41$	$80.13^{a} \pm 0.10$	$80.21^{a} \pm 0.04$
Ash (per cent)	$1.34^{ab} \pm 0.05$	$1.40^{ab} \pm 0.02$	$1.37^{ab} \pm 0.01$	$1.37^{ab} \pm 0.01$	$1.38^{ab} \pm 0.06$	$1.43^{ab} \pm 0.01$	$1.54^{ab} \pm 0.04$
Total Fat (per cent)	$1.10^{a} \pm 0.10$	$1.36^{b} \pm 0.06$	$1.42^{b} \pm 0.02$	$1.26^{b} \pm 0.01$	$1.33^{b} \pm 0.01$	$1.33^{b} \pm 0.01$	$1.35^{b} \pm 0.15$
Crude Protein (per cent)	$15.20^{a} \pm 0.00$	$15.82^{a} \pm 0.43$	$16.97^{a} \pm 0.25$	$15.88^{a} \pm 0.26$	$15.78^{a} \pm 0.44$	$15.76^{a} \pm 0.24$	$15.43^{a} \pm 0.11$
Carbohydrates (per cent)	$0.04^{a} \pm 0.00$	$0.04^{a} \pm 0.22$	$0.09^{a} \pm 0.04$	$0.03^{a} \pm 0.25$	$0.07^{a} \pm 0.05$	$0.03^{a} \pm 0.02$	$0.17^{a} \pm 0.04$

Mean bearing similar superscript with small alphabet in the row do not differ significantly with each other.

Recommendation

Fish attained higher growth rate, good survival rate and specific growth rate in medium dose of chicken manure treated water. Proximate composition also shows that fishes treated in medium dose of chicken manure have comparatively higher protein and fat in muscles which support good growth and wellness of fish.

Benefit/Impact

Chicken manure alone was effective to stimulate productivity with conducive range of water quality and growth of fish without affecting proximate composition of fish meat.

89. Turkey Production under Intensive and Semi-Intensive Rearing System

Contact Scientist: **Dr. Asma Khan**

Methodology

Forty eight poults of eight weeks age were reared under intensive and semi-intensive system at Division of LPM, SKUAST-Jammu. Standard management practices were followed for rearing of birds. The experiment spanned over 90 days and body weight gain, feed consumption, feed conversion ratio, carcass characteristics and economy of production were recorded for both type of production system.

Table 1: Body Weight Gain (g) of Poults at different Weeks in different Rearing Systems

Age in Weeks	*Intensive System*	*Semi-Intensive System*
9	207.38 ±6.34[a]	158.33 ±7.91[b]
10	205.97±7.06[a]	141.19±9.61[b]
11	222.85±10.3[a]	187.61±9.28[b]
12	258.09±16.66[a]	181.42±10.16[b]
13	275.23±15.28[a]	217.61±11.77[b]
14	286.19±11.37[a]	219.68±12.67[b]
15	313.80±9.57[a]	219.76±13.30[b]
16	330.76±20.16[a]	210.47±12.63[b]
17	326.38±12.19[a]	218.57±13.32[b]
18	296.90±11.42[a]	215.23±10.63[b]
19	314.38±13.56[a]	220.47±12.71[b]
20	332.04±12.47[a]	238.14±14.56[b]
AWG	280.83±13.65	203.88±8.79

Mean bearing similar superscript with small alphabet in the row do not differ significantly with each other.

Table 2: Carcass Traits of Turkeys under different Systems of Rearing

Carcass Traits	*Intensive System*	*Semi-Intensive System*
Live weight (g)	4323.33±82.71	3372.50±117.35
Bled Weight (per cent)	95.90±0.67	95.10±1.29

Carcass Traits	*Intensive System*	*Semi-Intensive System*
Defeathered weight (per cent)	90.60±1.35	90.65±1.92
Carcass weight (per cent)	64.89±2.06	65.05±1.98
Head weight (per cent)	3.36±0.31	3.02±0.32
Shank weight (per cent)	3.59±0.34	3.24±0.36

Recommendation

Intensive type of production system is ideal for farmers who want faster body weight gain of birds and have less space. Semi-intensive type of turkey production is more economical and recommended for farmers having scavenging spaces.

Benefit/Impact

Turkey production is an economical enterprise and could boast the income of farmers.

Division of Veterinary Surgery and Radiology

90. Balanced Total Intravenous Anaesthesia Protocols for Adult Mules

Contact Scientist: **Dr. M.S. Bhadwal**

Introduction

Under ideal conditions, the anaesthesia should be induced in premedicated equines with injectable drugs and maintained with inhalant anaesthetics, but the facility of inhalant anaesthesia in large animals is not available even in most of the veterinary colleges in India, leave alone the field hospitals. Under such circumstances, the total intravenous anaesthesia (TIVA) that results in rapid and smooth induction and recovery; provides adequate analgesia and muscle relaxation, is desirable. The division has studied and standardized various TIVA combinations for short to long term surgery in adult mules.

Methodology

Various TIVA combinations have been standardized for short to long term surgery in adult mules. For all protocols the mules have to be examined for their health status and baseline parameters and fasted for 12-18 hours and water to be withheld for 6 hours.

Sites for jugular venipuncture have to be prepared for administration of various drugs.

A. Xylazine and Ketamine Combination

The xylazine hydrochloride is to be administered intravenously @ 1.3 mgkg^{-1}. Once the peak sedation indicated by lowering of head and ears, relaxation of lower lip, widening of base for support *etc.* is achieved (about 5 minutes after xylazine administration), anaesthesia can be induced by ketamine hydrochloride (2.5 mgkg^{-1}) intravenously. If padded induction area is available, the induction can be performed in standing animal and allowed to fall before restraining for surgery; otherwise the animal should be cast using standard methods after sedation and induction should be performed in recumbent animal. Caution is to be exercised to prevent forward fall of the animals.

With this protocol the induction is quick and smooth, quality of sedation, analgesia and recovery is moderate to good. However, the muscle relaxation is poor. The duration of anaesthesia is around 8.58±0.58 minutes.

B. Butorphanol, Xylazine, Ketamine Combination

In this anaesthetic protocol, premedication is done with butorphanol @ 0.04mg/kg intravenously, followed 30 min. later by xylazine @ 1.5 mg/kg intravenously. Anaesthesia is induced with ketamine @ 2.2 mg/kg intravenously.

The quality of anaesthesia is observed to be good with prolonged sedation and good analgesia with good muscle relaxation and smooth recovery. The duration of anaesthesia is observed to be only 14.21±2.0 minutes.

C. Butorphanol, Xylazine, Diazepam-Ketamine Combination

In this protocol Butorphanol tartrate should be administered IV @ 0.04 mgkg^{-1}, followed 30 min. later by xylazine @ 1.1 mgkg^{-1}, IV. At the time of peak sedation, following injection of xylazine, the induction is to be achieved with a diazepam-ketamine mixture in the ratio of 0.03 and 2.2 mgkg^{-1}, respectively, administered IV as in the previous protocol.

The anaesthesia is observed to last for a period of 19.17±1.95 min. without any complication under clinical trials. The quality of anaesthesia is good with prolonged sedation, good muscle relaxation and excellent analgesia, followed by a good and smooth recovery.

Recommendations

The TIVA using any of the above protocols should be used to achieve short term anaesthesia. The choice among the protocols should be based on the type of surgery, the muscle relaxation required and duration of the surgery. However; additional dose of ketamine or diazepam-ketamine mixture would be required for prolonging the anaesthesia in surgeries of longer duration.

Onset of sedation indicated by lowering of head, widening of support base

Ear drooping

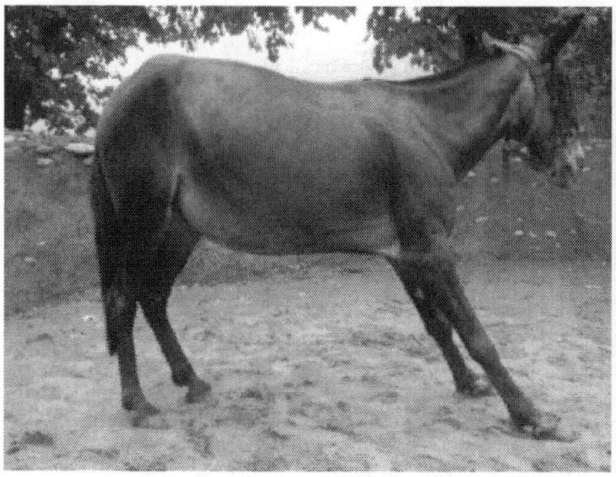

Wide stance of forelimbs to increase support base to avoid fall

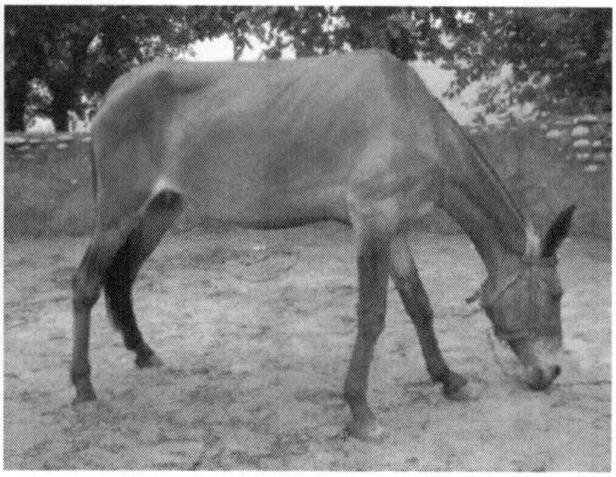

Wide stance of forelimbs with muzzle touching ground

Figure 1: Animals Showing Signs of Sedation following Butorphanol-Xylazine Administration.

Impact

Minor surgeries like castration, excision of small tumours, suturing of lacerated wounds, application of cast in fractures *etc.* can be performed at the field level by the local veterinarians, thereby saving the farmers from spending lot of money and time in transporting the animals for minor surgeries to the referral hospitals.

The anaesthetic protocols of TIVA are also routinely used at University TVCC, even for performing major abdominal surgeries, as there are no facilities for inhalant anaesthesia of large animals.

Mule in sternal recumbency following diazepam-ketamine administration

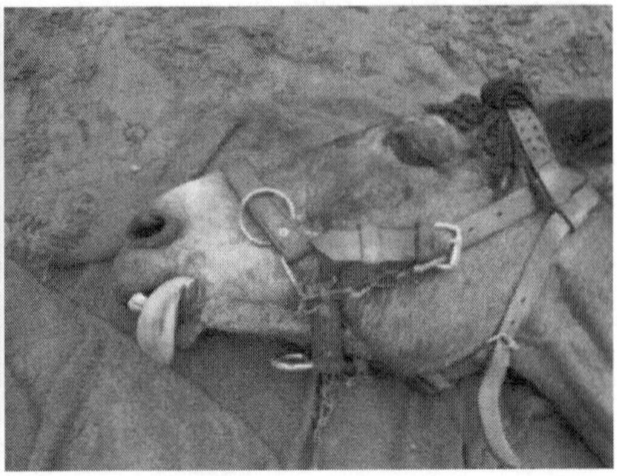

Relaxation of Tongue

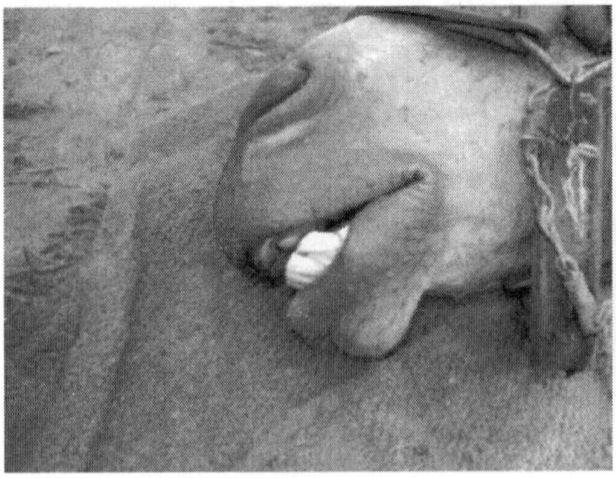

Relaxation of Jaw muscles

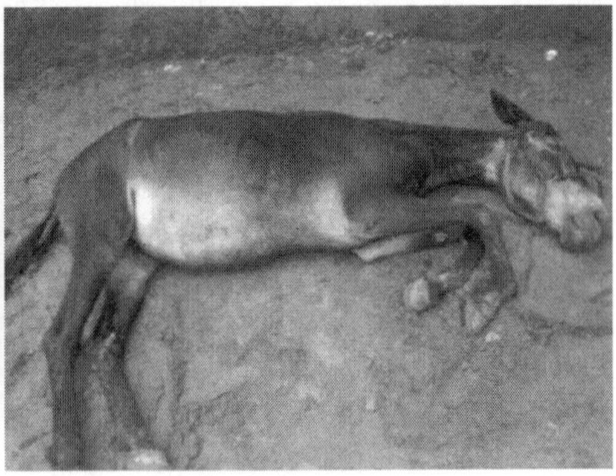

Animal in state of complete muscle relaxation during anaesthesia

Figure 2: Mules under Butorphanol-Xylazine-Diazepam-Ketamine Anaesthesia.

91. Management of Intussusception in Cattle

Contact Scientist: **Dr. H.R. Bharaddwaj**

Introduction

The intussusception can only be treated surgically, it has no medicinal treatment. The success rate of the surgery in literature varies from 30 per cent -50 per cent. However, the division, after several studies, has developed a protocol of surgical management that with preoperative and postoperative supplementation of intravenous fluid enriched with potassium and chloride results in success rate of >80 per cent.

Methodology

Once the intussusception is diagnosed, all the affected animals should undergo right flank laparotomy followed either by enterectomy (resection of the affected part of intestine) and intestinal anastomosis or by intestinal anastomosis bypassing the affected intestine under local anaesthesia in standing position. The anastomosis can be performed by end to end anastomosis, end to side anastomosis, side to side anastomosis. The anastomosis can be satisfactorily performed either manually by suturing or by stapling (using GIA stapling instrument, ILA 100, 3.8 x 4mm). However, Manual suturing (first layer of simple continuous pattern followed by second layer of Cushing pattern with Catgut "3-0") is routinely adopted for end to end enteroanastomosis.

Metabolic alkalosis and respiratory acidosis is a common finding in all cows suffering from intestinal obstruction.

Surgery followed by post-operative treatment with balanced mineral supplementation, polyionic fluid therapy, routine antibiotics (Enrofloxacin and Gentamicin) and analgesics is effective. Fluid therapy should be supplemented with potassium and chloride to counteract metabolic alkalosis and respiratory acidosis.

Recommendations

The history of colicky pain followed by cessation of defecation and anorexia in cattle should lead the veterinarian to the suspicion of intestinal obstruction due to colic. The condition can be confirmed by performing per rectal examination and palpation of a coiled mass of involved intestine with the feel of a sausage. However, in young animals where per rectal examination cannot be performed or in pregnant animals where the mass cannot be palpated due to gravid uterus occupying the caudal abdomen, the condition should be confirmed by ultrasonography.

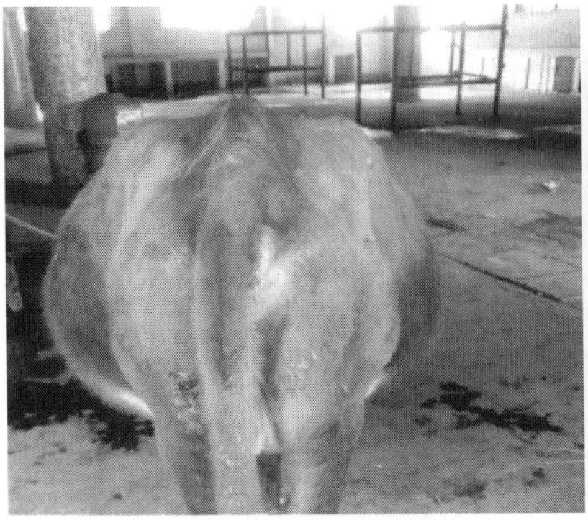

Severe abdominal distension in animal

Animal showing stretching of body (arching of back)

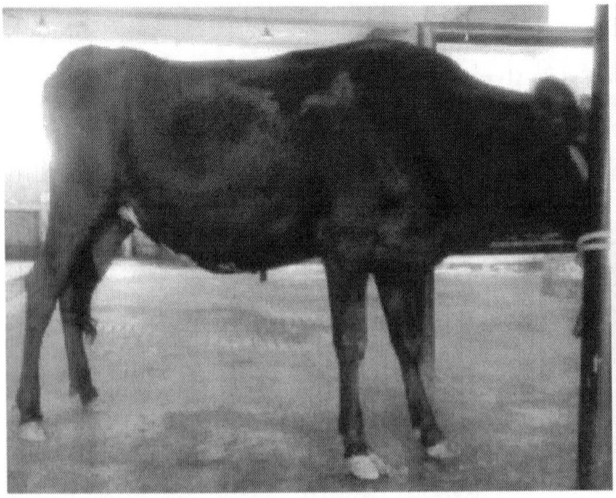

Animal standing with hind legs stretched out (sign of colic)

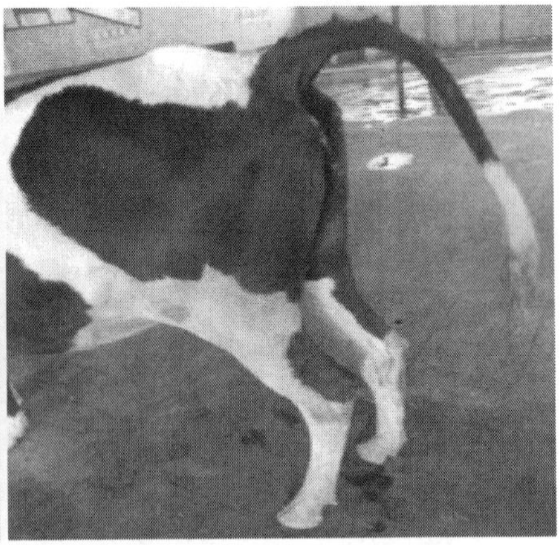

Futile attempts of defecation and tail lifting

Figure 1: Clinical Signs in Cattle Suffering from Intussusception.

Fluid therapy should be supplemented with potassium and chloride to counteract the electrolyte changes.

Antibiotics namely, Enfloxacin and Gentamicin are effective in preventing post surgical infections.

As the disease is seen in cattle exclusively fed wheat straw and wheat bran, balanced ration should be advised to minimize the occurrence of the disease. Sudden change of ration should be avoided and periodic deworming should be followed.

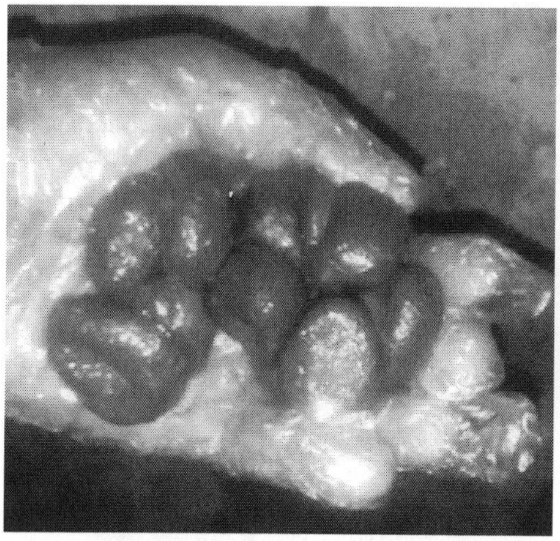

Firm green coloured faecal balls (seen in a fresh case) these are the last faeces present in rectum

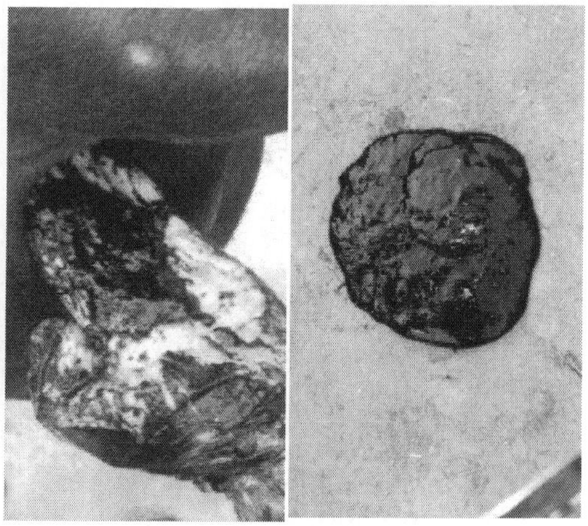

Tarry coloured contents in the rectum

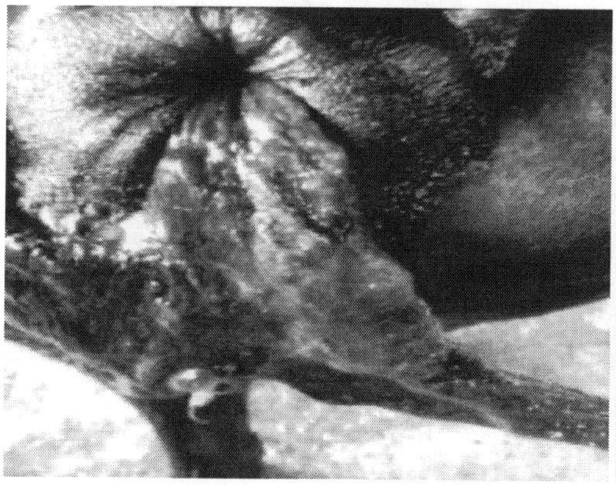

Blood and mucus mixed contents (contents are not tarry coloured)

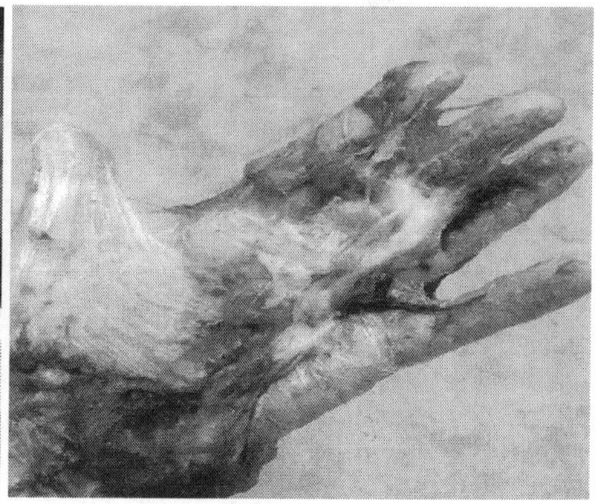

contents

Figure 2: Contents on Rectal Examination.

Impact

The recommendations will improve the efficiency of surgeons and aid in providing targeted treatment to the affected animals, thereby benefitting the farmers and other animal owners indirectly by saving their precious animals.

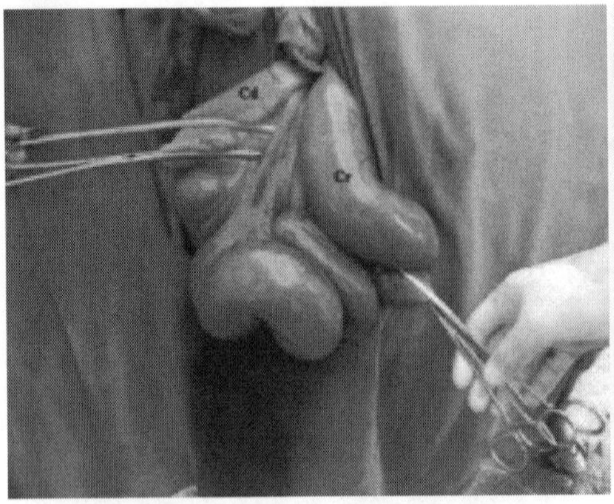

Cranial (Cr) and caudal (Cd) ends of intussusception mass

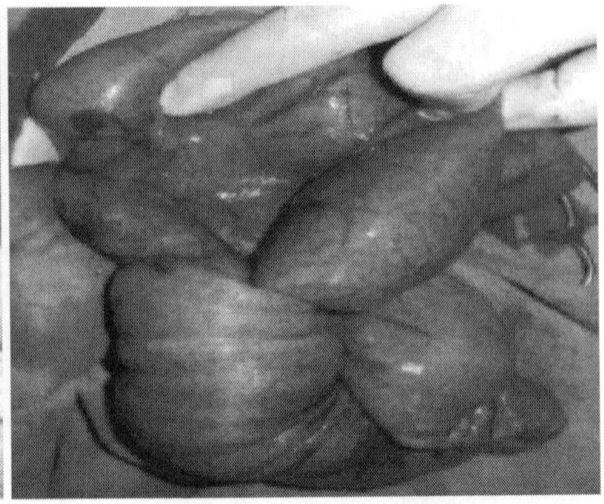

Telescoping point of intussusception

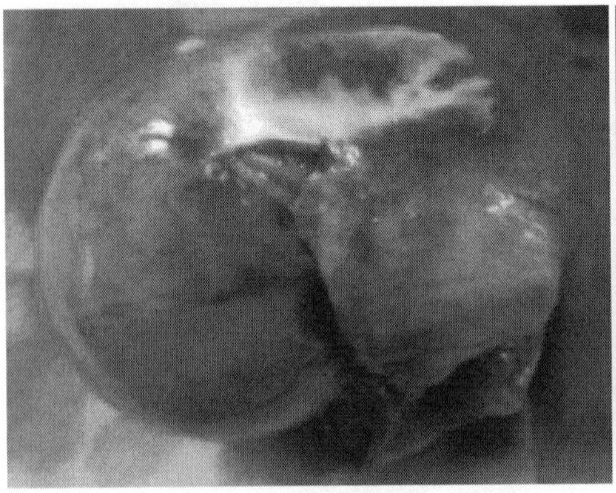

Resected intestine.
White fibrin depict the adhesions

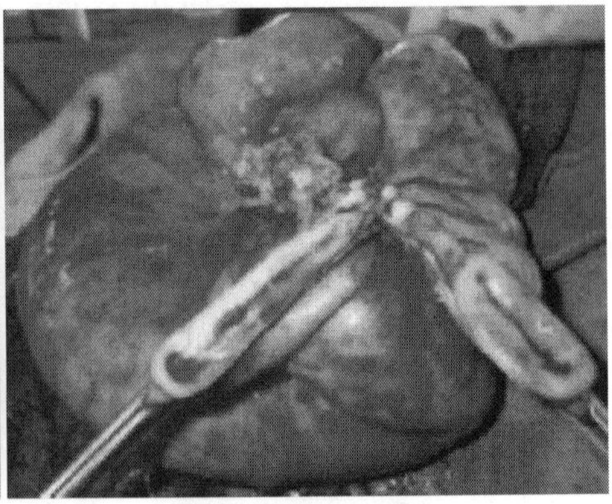

Intestinal ends before anastomosis

Intestine after anastomosis

Figure 3: Intraoperative Observations.

92. Post Operative Pain Management in Dogs

***Contact Scientist:* Dr. M.S. Bhadwal**

Introduction

"The relief of animal suffering," the concept has gained importance in the society, as also among animal health professionals worldwide. Pain management improves the recovery process, whether from illness, surgery, or injury. Best of all, because it reduces stress and increases a sense of well being, pain management may even help your furry friend live longer (American Association of Feline Practitioners/American Animal Hospital Association (AAFP/AAHA), 2007). The division conducted clinical studies on female dogs undergoing ovariohysterectomy to evaluate various drugs for post operative analgesia. post-operative pain assessment was made using three pain scales *viz.*, Melbourne Pain Scale and Dynamic and Interactive Visual Analogue Scale and Glasgow Pain Scale at 2, 4, 8, 12 and 24 hours post-operatively.

Methodology

The dogs were premedicated with atropine (0.04 mg/kg, SC) and diazepam (0.5mg/dl). In all the dogs, ovariohysterectomy was performed through mid-ventral incision under general anaesthesia induced and maintained with 5 per cent thiopentone sodium. All the analgesic drugs were administered immediately after completion of surgery.

Carprofen

Carprofen was administered @ 4mg/kg, IV as a single dose.

Meloxicam

Meloxicam is administered @ 0.2mg/kg, IV as a single dose.

Tramadol

Tramadol is administered @2mg/kg, IM, repeated at same dose rate after every six hours.

Nimesulide

Nimesulide is administered @4mg/kg, IM, as a single dose.

Combination of Tramadol and Nimesulide

Tramadol (2mg/kg, IM) and Nimesulide (4mg/kg, IM) in combination.

Combination of Tramadol and Nimesulide

Tramadol (1mg/kg, IM) and nimesulide (2 mg/kg, IM)

It has been observed that the combination of the drugs is better than the use of the drugs alone in controlling pain induced by ovariohysterectomy.

Recommendations

Although the above drugs and their combinations have been found to be effective in controlling the post opertaive pain in ovario-hysterectomised dogs, it is better to use combination of drugs at lower dose rates. All animals undergoing any surgery or suffering from trauma must be provided suitable analgesics.

Impact

It is inhuman and unethical to perform surgery without providing post operative analgesia. When recovering from invasive procedures, animals may not only be in pain, but also weak and disoriented. The post operative pain relief not only reduces post-operative suffering in animals, but also improves recovery from anaesthesia.

93. Ultrasound Guided and Laparoscopic Needle Biopsy in Dogs

Contact Scientist: **Dr. A.K. Gupta**

Introduction

Every veterinarian attending a clinical case is faced with the challenge of arriving at a definitive diagnosis, so as to formulate an appropriate treatment regimen and give a prognosis. The spectrum of diagnosis can be narrowed down to a particular system and organ based on history, symptoms, clinical examination, haemato-biochemical alterations and various imaging techniques, but definitive diagnosis, especially in diseases involving abdominal organs, can only be reached at by histopathologic examination of a tissue sample from the suspected organ. To be clinically useful, biopsy sample should be large enough for a pathologist to accurately characterize any lesion that might be present. In addition crushing and fragmentation of the specimens should be minimized. Biopsy can be obtained by a number of techniques, but with the availability of ultrasound and laparoscopy with which the biopsy can be obtained under their guidance; there is little justification of using other invasive or blind methods. The division has standardized the procedure of obtaining needle biopsy of internal organs of dogs under ultrasound guidance and also using laparoscopy to aid diagnosis.

Methodology

Double spring loaded automatic 16-18 gauge, 16 cm long disposable biopsy instrument with a throw of 22mm and sample notch of 19mm), is used.

Ultarsound Guided Biopsy

The ultrasound guided biopsy of liver, spleen and prostate is carried out in dorsal recumbency, whereas for the renal biopsy some animals require to be tilted 30-45 per cent to the contra-lateral side. Based on preliminary ultrasonographic examination, the organ and site for biopsy isselected; skin is prepared by scrubbing with 0.25 per cent Chlorehexidine-alcohol solution. The transducer is cleaned with the same solution. Sterilized coupling gel is applied liberally on the skin; the biopsy site is identified and biopsy needle introduced close to one end of the transducer at an angle almost 30°- 45° to the long axis of the transducer and maintaining the needle at the centre of the transducer head to remain within the scan beam. The biopsy gun is loaded before insertion and biopsy performed with free hand technique by pressing the trigger as soon as the needle tip is seen just penetrating the capsule of the organ or enters the lesion. The biopsy procedure is performed by a single examiner with the right hand while holding the transducer with the left hand. The biopsy sample is immediately transferred to 10 per cent neutral buffered formalin (NBF), by gently shaking the biopsy needle in the NBF, for histopathologic processing. The biopsy site is observed for 2-3 minutes for haemorrage, if any.

Laparoscopic Beedle Biopsy

Laparoscopy can be performed using ventral midline approach under general anaesthesia. With the dog in dorsal recumbency, primary port, for telescope, is placed on the midline, 1 to 1.5 inches caudal to the umbilicus. Depending upon the organ to be biopsied the secondary port, for forceps or for palpation probe, is to be placed about 2 to 3 inches cranio-lateral to the primary port on left or right side for liver, spleen and kidneys, whereas for approaching the prostate the secondary port is placed about 2 inches caudo-lateral to the primary port.

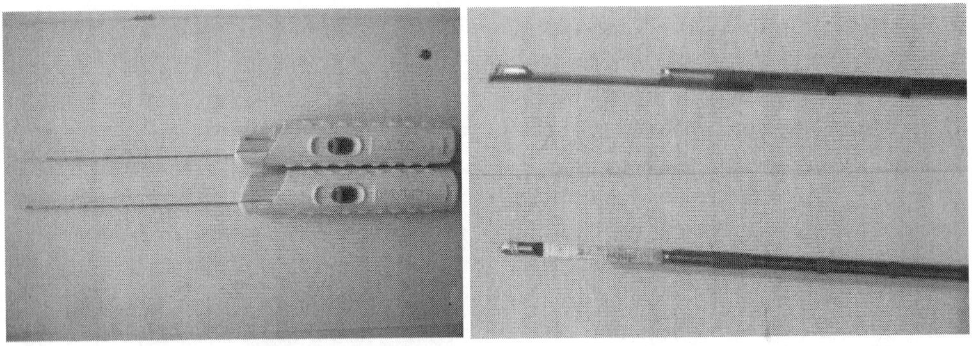

Biopsy gun and laparoscopic equipment

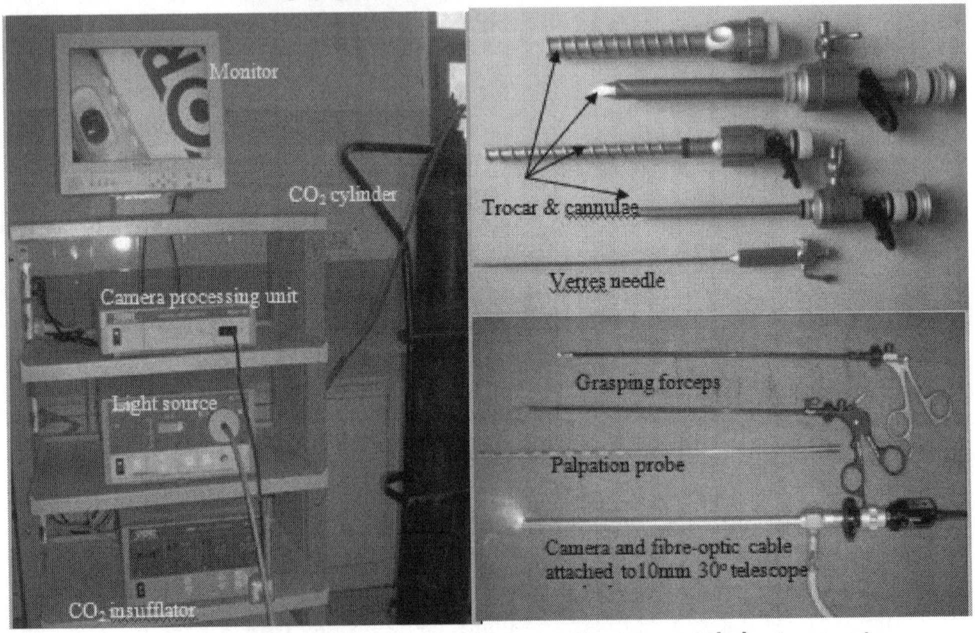

Laparoscopic unit **Laparoscopic instruments**

Figure 1: Equipment for Laparoscopic Biopsy.

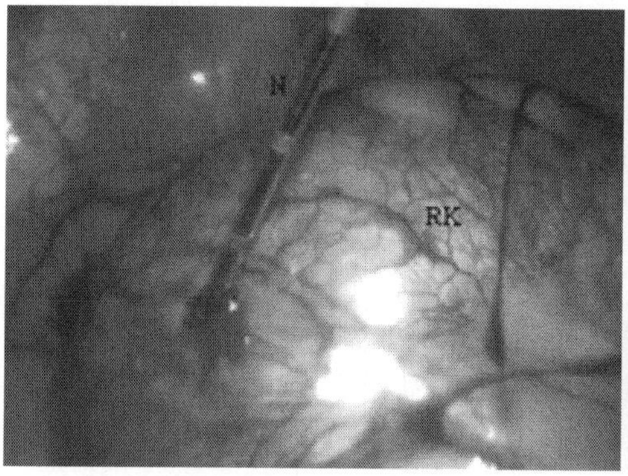

Biopsy needle (N) has penetrated cortex of caudal pole of right kidney (RK) and cut the biopsy tissue.

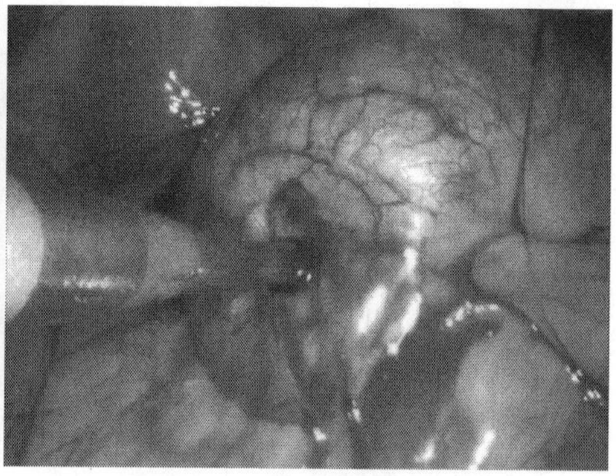

Laparoscopic view showing application of pressure with palpation probe to achieve haemostasis following right renal biopsy.

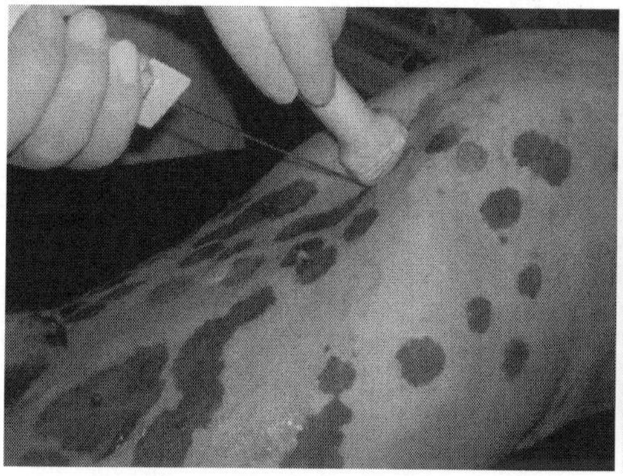

Position of biopsy needle and ultrasound transducer for obtaining liver biopsy (Sagittal plane).

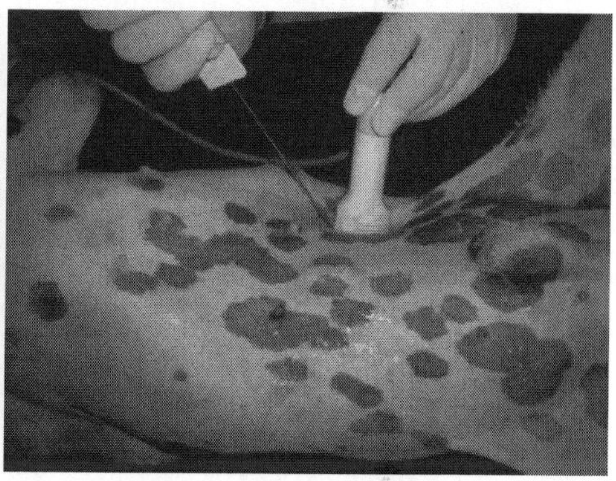

Position of biopsy needle and ultrasound transducer for obtaining biopsy from caudal pole of left kidney (Sagittal plane).

Figure 2: Laparoscopic and Ultrasound Guided Biopsy in Dogs.

The site of insertion of the biopsy needle varies and is selected by pressing the abdominal wall close to the organ/lesion of interest and the biopsy needle is inserted under direct visualization and when the tip depresses the capsule of the organ slightly, it is operated to cut the biopsy. The needle is removed and sample transferred to NBF. The biopsy site is observed for bleeding, if any and stopped by application of pressure with palpation probe for 1-2 minutes, whenever needed.

Recommendations

If the facilities are available both the techniques of obtaining biopsy are recommended. Both have their own merits and demerits. Ultrasound guidance helps identification and biopsy of deep seated parenchymal lesions, whereas laparoscopic visual guidance helps direct visualization of the surface of all abdominal organs and bleeding, if any, can be easily controlled. Ultrasound guided biopsy can be obtained under general anaesthesia of under local analgesia, whereas laparoscopy requires general anaesthesia. Laparoscopic equipment is expensive when compared with ultrasound machine. Both the techniques are safe when performed by an expert.

Impact

Both, the ultrasound guided and laparoscopic biopsies of internal organs have been of immense help in arriving at confirmatory diagnosis in the hospital of TVCC of the university where both the facilities are available.

94. Ultrasound Guided Biopsy of Liver and Kidney in Cattle

Contact Scientist: **Dr. M.S. Bhadwal**

Introduction

Accurate diagnosis is the most important step in the management of a disease condition. The history, clinical examination, urinalysis, haematology and blood biochemistry only lead us to diagnose the organ or system affected, but it is the histopathology of the biopsy that helps in differentiating various disease conditions affecting the organ or system. The division has standardized the technique of obtaining biopsy under ultrasound guidance that is accurate and minimally invasive.

Methodology

Ultrasonography of liver and kidneys in cattle is performed with 3.5 to 7.5 MHz micro-convex and linear transducers. Ultrasound guided biopsy of liver and kidneys can be easily obtained using 18G automated biopsy needle.

Liver

The last three intercostal spaces on the right side are best suited for ultrasonography of liver. The site is shaved and prepared as for aseptic surgery. Sterilized coupling gel is applied liberally and entire liver is scanned. The hepatic parenchyma appears as numerous weak echoes distributed homogenously. The size of liver is greatest in 10th intercostal space. The caudal vena cava, portal vein and gall bladder are observed to have a characteristic triangular, circular and pear shapes, respectively, on sonograms. Studies on cattle have revealed that the distance between the ventral liver margin and the dorsal midline of the back is observed to be comparatively shorter in the 12th intercostal space. The depth of caudal vena cava and portal vein is greater in 11th and 12th intercostal space, respectively.

Any lesion in the liver is identified and then liver biopsy is obtained easily through the 11th intercostal space under ultrasound guidance using free hand technique keeping the tract of the needle within the field of view.

Kidney

Right kidney extends from 13th rib to 3rd lumbar vertebra in cattle. The renal artery, vein and ureter cannot be differentiated easily. Left kidney can be examined both trans-rectally and trans-abdominally. The kidney appears lobulated on ultrasonography. The renal cortex is hyperechoic than medullary pyramids. Biopsy of kidney can be performed at the upper right paralumbar fossa between the L2 and L3. No major complications are reported post biopsy.

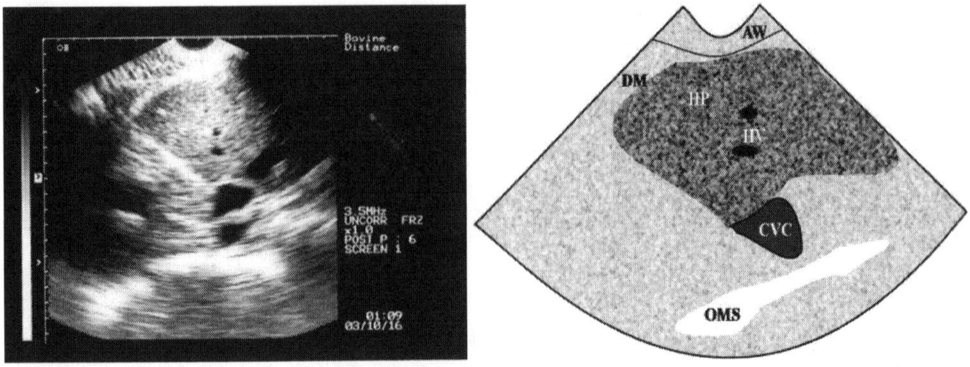

Figure 1: Sonogram and Schematic Representation of the Dorsal Margin of the Liver, Obtained from the 12th Intercostal Space, with 3.5 MHz Transducer; HP= hepatic parenchyma, DM= dorsal margin of liver, AW: abdomen wall, HV: hepatic vein, CVC: caudal vena cava, OMS: omasal wall.

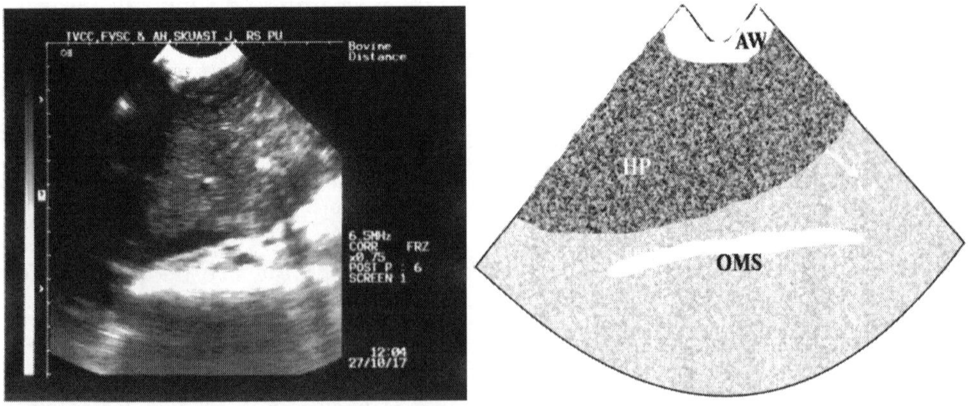

Figure 2: Sonogram and Schematic Representation of the Normal Liver Texture, Obtained from the 11th Intercostal Space with 6.5 MHz Transducer, Showing the Ventral Border of the Liver; HP: hepatic parenchyma, AW: abdomen wall.

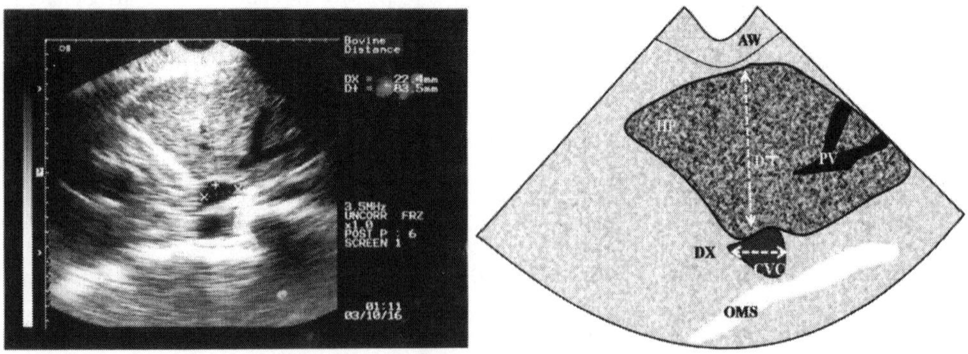

Figure 3: Sonogram and Schematic Representation of the Normal Liver Texture, Portal Vein and Caudal Vena Cava, Obtained from the 12th Intercostal Space with 3.5 MHz Transducer; HP: hepatic parenchyma, AW: abdomen wall, CVC: caudal vena cava, DX: diameter of CVC, D+: thickness of liver over CVC, PV: portal vein, OMS: omasal wall.

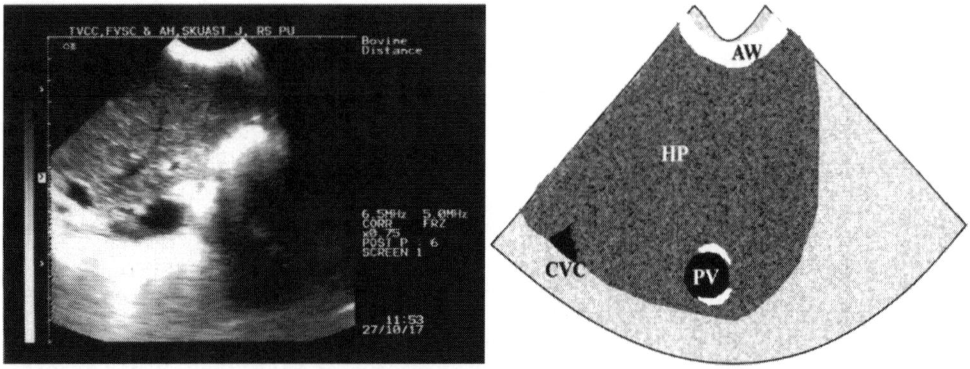

Figure 4: Sonogram and Schematic Representation of Caudal Vena Cava and Portal Vein, Obtained from the 11th Intercostal Space with 6.5 MHz (frequency adjusted to 5 MHz) Transducer, Showing the Characteristic Triangular and Circular Shapes of CVC and PV in Cross Section, Respectively; HP: hepatic parenchyma, AW: abdomen wall, CVC: caudal vena cava, PV: portal vein.

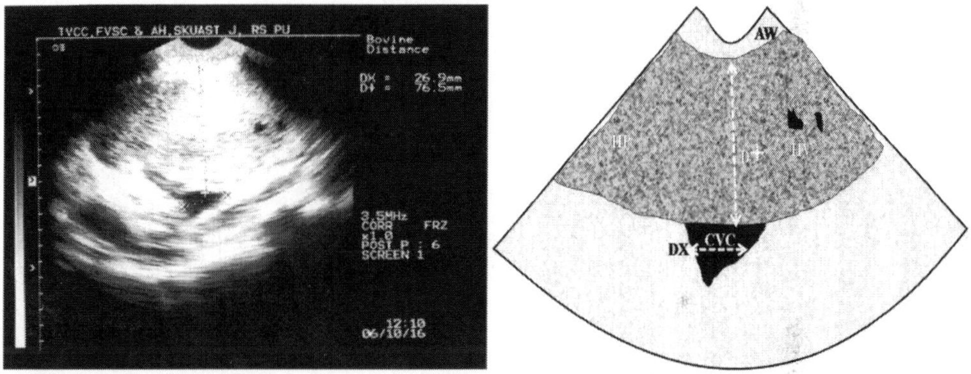

Figure 5: Sonogram and Schematic Representation of the Normal Liver Texture and Caudal Vena Cava, Obtained from the 11th Intercostal Space with 3.5 MHz Transducer; HP: hepatic parenchyma, AW: abdomen wall, CVC: caudal vena cava, DX: diameter of CVC, D+: thickness of liver over CVC, HV: hepatic vein.

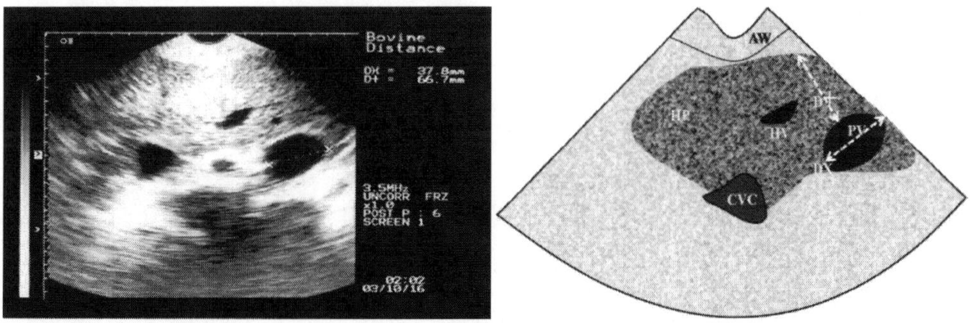

Figure 6: Sonogram and Schematic Representation of the Normal Liver Texture and Portal Vein, Obtained from the 12th Intercostal Space with 3.5 MHz Transducer; HP: hepatic parenchyma, AW: abdomen wall, CVC: caudal vena cava, PV: portal vein, DX: diameter of PV, D+: depth of PV, HV: hepatic vein.

Recommendations

The ultrasonography is a reliable non-invasive technique for evaluation of the liver and kidneys in cattle which has been standardized for the collection of biopsy samples from the clinical cases. It is recommended for use in field by veterinarians where ultrasonography machines are available.

Impact

Histopathology is required for arriving at confirmatory diagnosis of diseases of parenchymatous organs. Obtaining biopsy under ultrasound guidance is a simple and safe procedure. It helps in confirming the disease diagnosis so that specific treatment can be instituted. As of now, the technique is not used by the field veterinarians due to lack of the ultrasound facility, but is routinely used on clinical cases in the University TVCC.

Division of Livestock Products Technology

95. Development of Fiber Enriched Chicken Sausages

Contact Scientist: **Dr. Sunil Kumar**

With increasing awareness about the effects of the foods on the health of humans, there has been a continuous increase in the demand of health based foods which cater the special needs of the people. Fiber fortification is one such area which has received considerable attention during last decade. Fiber enriched meat products have been developed by the division by utilizing locally available fiber sources *i.e.* kohlrabi, pumpkin and carrot.

Methods

Lean meat from dressed chicken was cut into smaller chunks and minced in a Sirman mincer (MOD-TC 32 R10 U.P. INOX, Marsango, Italy) with 6mm plate. The common salt, vegetable oil, refined wheat flour (maida), nitrite, sodium tripolyphosphate, spice mixture and condiment mixture were added to weigh meat according to formulation. Meat emulsion for chicken *sausages* was prepared in Sirman Bowl Chopper [MOD C 15 2.8G 4.0 HP, Marsango, Italy]. Minced meat was blended with salt, sodium tripolyphosphate and sodium nitrite for 1.5 minute. Water in the form of crushed ice was added and blending continued for 1 minute.

This was followed by addition of refined vegetable oil and blended for another 1 to 2 minutes. This was followed by addition of condiments, spice mixture, eggs and other ingredients (vegetable fibre according to the requirement of experiment) and again mixed for 1.5 to 2 minutes to get the desired emulsion. Adequate care was taken to keep the end point temperature below 18°C by preparing the emulsion in cool hours of morning, by addition of meat and other ingredients in chilled/ partially thawed form and by addition of crushed ice or ice water. Formulation of chicken *sausages* is presented in Table 3.

Formulation of *Sausages* from Chicken Meat

Ingredients	*Per cent (W/W)*
Lean meat	64.45
Added water	10.0
Vegetable oil	10.0
Condiment mixture	5.0
Refined wheat flour	3.0
Spice mixture	2.0
Table salt	1.75
Monosodium Glutamate	0.5
Sodium Tripolyphosphate	0.3
Egg	3.0
Sodium nitrite	120 ppm

Flow Diagram for Preparation of *Chicken Sausages*

Dressed chicken

↓

Removal of skin and fat

↓

Deboning

↓

Cutting of deboned meat into smaller chunks

↓

Mincing through 6mm plate

↓

Addition of salt (1.75 per cent), STPP (0.3 per cent) and sodium nitrite (120 ppm)

↓

Chopping for 1.5 minute

↓

Addition of water (10 per cent) in the form of crushed ice

↓

Chopping for 1 minute

↓

Addition of vegetable oil (10 per cent)

↓

Chopping for 1-2 minute

↓

Addition of spice mixture, condiments mixture and refined wheat flour

↓

Chopping for 1.5-2 minutes

↓

Emulsion

↓

Stuffing of emulsion in casings

↓

Cooking in Hot air oven (140±5°C/25 mins)

↓

Packaging and Storage at refrigeration temperature

Stuffing of *Sausages*

The Emulsion was filled in to the Sirman sausage filler (Model-1S-V15-IRDA-VERT, S.No. 07L01410. Marsango, Italy), and the artificial cellulose casing was applied on the nozzle of sausage filler. Pressure was applied in such a way so that the emulsion was filled in to the casings.

Cooking of *Sausages*

Hot air oven method was used for cooking of *sausages.*

Hot-Air Oven Method

Raw *sausages* were hanged vertically on skewers which were placed at the two ends of hot air oven. (Yorco sales Pvt. Ltd. India, Model-YS1-431, S. No. 02B2843). Raw *sausages* were cooked in a preheated hot air oven at 140±5°C for a total time of about 25 minutes. The internal temperature of *sausages* was monitored by a thermometer and cooked to an internal temperature of 80±2°C. The *sausages* were removed from the skewers, cooled to room temperature and weighed. Pooled sample of each treatment was assigned for analysis.

Result

Table 1: Effect of Optimum Levels of Kohlrabi, Pumpkin, and Carrot on pH, Cooking Yield and Proximate Composition of Cooked Chicken *Sausages* (Mean± SE)*

Parameters	*Levels of Vegetable Fibre (per cent)*			
	0	*Carrot 12 per cent*	*Pumpkin 12 per cent*	*Kohlrabi 6 per cent*
pH	6.33^{c} ±0.02	6.23^{b} ±0.01	6.25^{b} ±0.01	6.11^{a} ±0.02
Cooking Yield (per cent)	92.11^{c} ±0.26	89.62^{a} ±0.16	91.21^{b} ±0.13	91.25^{b} ±0.07
Moisture (per cent)	63.23^{a} ±0.30	66.15^{bc} ±0.27	66.83^{c} ±0.28	65.41^{b} ±0.12
Crude Protein per cent)	17.06^{b} ±0.27	16.05^{a} ±0.05	16.17^{a} ±0.01	17.08^{b} ±0.03

Parameters	*Levels of Vegetable Fibre (per cent)*			
	0	*Carrot 12 per cent*	*Pumpkin 12 per cent*	*Kohlrabi 6 per cent*
Ether extract (per cent)	13.57^{b} ±0.31	12.59^{a} ±0.18	13.57^{b} ±0.02	13.40^{b} ±0.04
Crude Fibre (per cent)	0.73^{a} ±0.04	1.01^{bc} ±0.06	1.13^{c} ±0.03	0.88^{b} ±0.01
Ash (per cent)	2.82^{c} ±0.02	2.47^{a} ±0.02	2.49^{a} ±0.03	2.66^{b} ±0.03

*Mean ±SE with different superscripts in a row differs significantly ($P<0.05$). n = 6.

Recommendation

Fiber enriched chicken sausages can be developed by incorporating 12 per cent pumpkin or 12 per cent carrot or 6 per cent kohlrabi replacing lean meat in the formulation without affecting the sensory characteristics of the products.

Benefit/Impact

The products developed have significantly higher dietary fiber besides the good quality animal proteins. The developed products have a shelf life of 14 days under refrigerated environment within low density polyethylene pouches.

96. Shelf-Stable Fish Snacks

Contact Scientist: Dr. Sunil Kumar

Snack foods are foods for light meals eaten in between meals and during relaxation for pleasure. Snack food is one of the fastest growing segment of the food industry. Snacks can be prepared on small scale entrepreneurs by youth and women co-operatives without any need of complicated machinery. Since snacks foods need no refrigeration and have extended shelf life this allows limitless marketing opportunities.

Method

Methodology for the preparation of fish *snacks* is described in following steps *i.e.* preparation of emulsion, filling in manual extruder, cooking of fish *snacks*.

Preparation of Emulsion

Several preliminary trials were conducted to standardize the formulation for preparation of emulsion for fish *snack* as indicated in Table mentioned below. Fish meat was cut into smaller chunks and minced in a Sirman mincer (MOD-TC 32 R10 U.P. INOX, Marsango, Italy) with 6mm plate. The common salt, refined wheat flour (maida), sodium nitrite, sodium tripolyphosphate, sodium bicarbonate, spice mixture and condiment mixture were added to weighed meat according to formulation. Meat emulsion for fish was prepared in Sirman Bowl Chopper [MOD C 15 2.8G 4.0 HP, Marsango, Italy]. Minced meat was chopped with all the curing ingredients for 1.5 minute. Water in the form of crushed ice was added and blending continued for 1 minute. This was followed by addition of spice mixture, condiments and other ingredients and again mixed for 1.5 to 2 minutes to get the desired emulsion. Adequate care was taken to keep the end point temperature below 18°C by preparing the emulsion in cool hours of morning, by addition of meat and other ingredients in chilled/partially thawed form and by addition of crushed ice or ice water. Formulation of fish *snacks* is presented in following table.

Filling of Manual Extruder

Emulsion obtained was filled into the manual extruder and by manual rotation of its handle, *snacks* of desired shape were obtained.

Cooking of Fish *Snacks*

Depending upon the experimental requirement, frying method was employed for cooking of fish *snacks*. *Snacks* obtained from the manual extruder were directly fried in the refined cottonseed oil.

Formulation for Fish *Snacks*

Ingredients	*Per cent (W/W)*
Fish meat	47
Added water	10
Condiment mixture	2.7
Rice flour	30
Refined wheat flour	5
Spice mixture	2
Table salt	2.5
Sodium tripolyphosphate	0.3
Sodium nitrite	150ppm
Sodium bicarbonate	0.5

Flow Diagram for Preparation of Fish *Snacks*

Fish

↓

Removal of scales and fins/evisceration

↓

Deboning

↓

Cutting deboned meat into smaller chunks

↓

Mincing through 6mm plate

↓

Addition of salt, STPP, nitrite and sodium bicarbonate

↓

Chopping for 1.5 minute

↓

Addition of water (10 per cent) in the form of crushed ice

↓

Chopping for 1 minute

↓

Addition of spice mixture, condiments mixture and refined wheat flour, Rice flour

↓

Chopping for 1.5-2 minutes

↓

Emulsion/*snack* mixture

↓

Filling into manual extruder

↓

Cooking in oil/deep fat frying

↓

To bring snacks at room temperature

↓

Packaging

Table 1: Effect of Ambient Temperature Storage on Sensory Attributes of Aerobically Packaged Cooked Fish *Snacks* Incorporated with Rice Flour (RF), Corn Flour (CF), Blackgram Flour (BGF), Peanut Flour (PNF) (Mean± SE)*

Treatments	Storage Period (Days)				
	0	7	14	21	28
Colour and appearance					
RF (30 per cent)	6.28 ±0.11^{Ac}	6.21 ± 0.11^{Ac}	6.02± 0.08^{Ac}	5.78±0.12 ^{Ab}	5.26 ±0.17^{a}
CF (20 per cent)	6.88 ±0.06^{Bc}	6.78 ± 0.08^{Bc}	6.57± 0.10^{Ac}	6.28 ±0.12^{Ab}	5.21± 0.16^{a}
BGF (10 per cent)	7.45± 0.07^{Cc}	7.33 ± 0.07^{Cc}	6.84 ±0.10^{Bc}	6.73 ±0.13^{Bb}	5.30± 0.20^{a}
PNF (10 per cent)	7.50± 0.09^{Cc}	7.42 ± 0.09^{Cc}	7.23±0.05 ^{Bc}	6.83 ±0.07^{Bb}	5.45 ±0.10$_{a}$
Flavour					
RF (30 per cent)	6.42 ±0.11^{Ac}	6.26 ±0.13^{Ac}	6.19 ±0.09^{Ac}	5.57 ±0.11^{Ab}	4.57 ± 0.13^{a}
CF (20 per cent)	6.78 ± 0.11^{Bc}	6.54 ± 0.10^{Ac}	6.45 ±0.09^{Ac}	5.61± 0.10^{Ab}	4.76 ± 0.17^{a}
BGF (10 per cent)	7.19 ± 0.08^{Cd}	6.97 ±0.10^{Ccd}	6.78 ± 0.11^{Bc}	5.52 ±0.11^{Ab}	4.30 ± 0.22^{a}
PNF (10 per cent)	7.23 ± 0.06^{Cc}	7.11 ± 0.06$_{Cc}$	6.90 ±0.07^{Bc}	5.92 ±0.09^{Bb}	4.80 ± 0.17^{a}
Crispiness					
RF (30 per cent)	6.26 ±0.13 ^{Ac}	6.19± 0.12^{Ac}	5.95±0.09$_{Ac}$	5.47±0.012^{b}	4.59 ± 0.11^{a}
CF (20 per cent)	7.19 ±0.08^{Bd}	7.09 ± 0.09^{Bd}	6.51±0.10$_{Bb}$	5.92±0.14^{b}	4.66 ± 0.15^{a}
BGF (10 per cent)	7.04 ±0.04$_{Bd}$	6.95 ± 0.05^{Bd}	6.58±0.11$_{Ac}$	5.97±0.90^{b}	4.76 ± 0.15^{a}
PNF (10 per cent)	7.16 ±0.09$_{Bd}$	7.09±0.05^{Bd}	6.76±0.08$_{Bc}$	6.07±0.11^{b}	4.88 ± 0.10^{a}
Texture					
RF (30 per cent)	6.83 ± 0.12^{BCcd}	6.64±0.12^{Acd}	6.42±0.11^{Ab}	6.28±0.11^{b}	5.64±0.11^{a}
CF (20 per cent)	6.47±0.09^{Ac}	6.40±0.08^{Abc}	6.30±0.08^{Abc}	6.11±0.11^{b}	5.76±0.11^{a}
BGF (10 per cent)	7.09±0.09^{Cc}	7.02±0.08^{Bc}	6.78±0.10^{Bc}	6.30±0.15^{b}	5.64±0.10^{a}
PNF (10 per cent)	6.76 ± 0.10^{ABc}	6.69±0.10^{Ac}	6.54 ± 0.10^{ABbc}	6.33±0.12^{b}	5.97±0.11^{a}
Overall acceptability					
RF (30 per cent)	6.35±0.10^{Ac}	6.26 ± 0.11^{Ac}	6.07±0.10 ^{Ac}	5.69±0.13^{b}	4.28±0.14^{a}
CF (20 per cent)	7.11±0.08^{Bc}	6.97 ± 0.12^{Bc}	6.64 ± 0.11^{Bc}	5.92 ± 0.07^{b}	4.59±0.19^{a}
BGF (10 per cent)	7.26 ±0.07^{BCd}	7.14 ±0.08^{Bc}	6.73 ±0.09^{Bc}	5.76 ±0.16 ^{b}	4.23±0.16^{a}
PNF (10 per cent)	7.45 ±0.09^{Cc}	7.26 ± 0.08^{Bc}	6.88 ±0.08^{Cc}	5.95 ±0.11^{b}	4.71±0.12^{a}

*Mean± SE with different superscripts in a row differs significantly ($p<0.05$).Mean values are scores on 8 point descriptive scale where 1- extremely poor and 8- extremely desirable

Recommendation

Shelf-stable fish snacks can be developed by incorporating three flours *viz.* 20

per cent corn flour, 10 per cent black gram flour and 10 per cent peanut flour by replacing rice flour (30 per cent) in the basic formulation.

Benefit/Impact

The snacks developed can be conveniently packed in LDPE pouches for a period of 21 days without any marked loss of physicochemical, microbial and sensory quality at ambient temperature. Fish-flour based snacks can prove to be valuable items to the consumer as a source of essential amino acids and other nutrients besides providing a scope of earning.

97. Designer Fiber-Enriched Chevon Cutlets

Contact Scientist: Dr. Sunil Kumar

To add to the variety of meat products, attempts were made to develop chevon cutlets. Further, the fiber content of the products was improved by incorporating different cereal fiber sources *i.e.* Quaker oats, Barley and Jowar flour to increase the functional value of the developed products.

Method

The ingredients used in chevon cutlets were lean goat meat, condiments, cooked shredded potato, crushed ice, salt, sugar, gram flour, whole egg liquid and spice mix. Lean meat from goat was cut into smaller chunks and minced in SIRMAN mincer (MOD-TC 32 R10 U.P. INOX, Marasango, ITALY) using 6 mm plate.

The minced goat meat, curing ingredients, table salt, crushed ice, sugar and sodium nitrite were added to minced chevon meat and this was kept for 15-20 min at refrigeration temperature (4±1°C). The mixture was shallow fat fried in 2.5 per cent w/w refined oil for 8 minutes. The condiments and spice mixture were fried separately till appearance of golden brown colour. The fried meat, fried condiments, gram flour, shredded potato and whole egg liquid were put into mixer. The batter so formed after blending was used for the preparation of cutlets by using moulds. After giving the shape, the cutlets were removed from mould and raw cutlets were kept at refrigeration temperature for 15-20 minutes for retaining shape. Now cutlets were dipped in whole egg liquid and rolled in rusk powder till uniform coating was formed on the surface and were deep fat fried in refined oil till they attained golden brown colour. The internal core temperature was measured with the help of a thermometer (80°C) and the excess fat was removed from the fried cutlets by using tissue paper. After attaining room temperature, the cutlets were packed in low density polyethylene pouches and were stored at refrigeration temperature (4±1°C).

Formulation of Chevon Cutlets

Ingredients	*Per cent (W/W)*
Lean meat	74.00
Crushed Ice	3.0
Sugar	0.25
Condiment mixture	10.0
Spice mixture	2.0
Table salt	1.75
Gram Flour	2.0
Shredded Potato	5.0
Whole Egg Liquid	2.0
Sodium nitrite	120 ppm

Flow Diagram for Preparation of Chevon Cutlets

Mincing of chevon meat

↓

Addition of common salt, sodium nitrite and sugar to minced meat and mix with mixer

↓

Store at refrigeration temperature for 15-20 minute

↓

Frying of minced cured meat in 2.5 per cent w/w refined soya bean oil

↓

Frying of condiments and spice mix separately

↓

Kneading of shredded potato, fried meat, fried condiments spice mix, gram flour and egg liquid

↓

Mixing with mixer grinder

↓

Moulding to cutlet shape

↓

Dipping in egg liquid

↓

Dusting with rusk powder

↓

Deep fat frying in refined vegetable oil (with internal core temperature 80°C or higher)

↓

Absorption of excess fat in tissue paper

↓

Cooling Packaging

Table 1: Effect of different Flours on the Sensory Attributes of Chevon Cutlets (Mean± SE)*

Attributes	*Flour Level (per cent)*			
	(CO) 0	*(OQ) 4*	*(BF) 4*	*(JF) 6*
Appearance and colour	7.00±0.06	7.14±0.06	7.07±0.05	7.04±0.04
Flavour	7.20±0.06[b]	7.12±0.05[ab]	6.98±0.05[a]	7.12±0.05[ab]
Juiciness	7.00±0.05	6.92±0.04	6.90±0.05	6.85±0.06
Texture	6.84±0.04	6.88±0.08	6.86±0.07	6.90±0.07
Overall acceptability	7.08±0.06	7.01±0.05	6.94±0.05	7.06±0.07

*Mean±SE with different superscripts in a row differ significantly ($P<0.05$). n = 21 for each treatment.

CO: Control; OQ: Oat quaker; BF: Barley flour; JF: Jowar flour,

Recommendation

Designer Chevon cutlets can be successfully prepared with the incorporation of 4 percent barley flour, 4 percent Quaker oats and 6 percent Jowar flour replacing lean meat.

Benefit/Impact

The storage quality of these fiber-enriched chevon cutlets can further be improved by using clove oil (100 ppm) as a natural preservative. The developed products can be stored successfully for 15 days under aerobic packaging under refrigerated conditions.

98. Development of Composite Meat Chocolate

Contact Scientist: Dr. Sunil Kumar

Composite meat chocolate, a novel product with enhanced nutritional quality was developed by using animal fat as vegetable fat replacer and then replacing a part of milk chocolate with dehydrated ground chicken meat. Calcium lactate as calcium supplement was added to the developed composite meat chocolate.

Method

For preparation of composite meat chocolate, first of all milk chocolate was prepared by thorough mixing of following ingredients, 7 – 11 per cent by weight of cocoa solids, 22 - 36 per cent by weight of cocoa butter, 6 – 12 per cent by weight of milk solids, 29 - 42 per cent by weight of sugar, 10 – 30 per cent by weight dehydrated chicken meat powder, (Write the exact composition in tabular form because it is not possible to give the range) 0.3 – 0.5 per cent emulsifier (being added after half of conching time is over) and ethyl vanillin in traces and said percentage being based on the total weight of the composite meat chocolate so as to produce an emulsion. Calcium lactate @ 1 per cent by weight of composite meat chocolate was added to this emulsion. The emulsion is then subjected to conching process for minimum 6 – 8 hours. Now the fruit extracts were added at desired level into the emulsion and let the conching be done for 30 minutes more. This is followed by tempering of the chocolate at 30 – 40°C for 15 – 20 minutes. Emulsion is now put into moulds of choice to obtain composite meat chocolate of desired shape. It was then wrapped in aluminium foil followed by LDPE packaging. The composition of composite meat chocolate is presented in table.

Formulation of Composite Meat Chocolate

Ingredient	*Percentage*
Cocoa solids	9.45 per cent
Cocoa butter	29.16 per cent
Animal fat	3.24 per cent
Milk Solids	9.9 per cent
Sugar	37.8 per cent
Lecithin	0.5 per cent 0.45
Ethyl vanillin	0.1 per cent (traces)
Dehydrated ground chicken meat	10 per cent
Calcium lactate	1 per cent of composite meat chocolate

Table 1: Effect of Calcium Lactate on Sensory Parameters of Composite Meat Chocolate (Mean± SE)*

Parameter	*C 2*	*C 3*	*T 1*	*T 2*	*T 3*
Appearance	8.54 ±0.20	8.55 ±0.20	8.55 ± 0.14	8.54 ± 0.21	8.54 ± 0.18
Taste	8.22 ±0.21^{b}	8.19 ± 0.21^{b}	8.03 ± 0.28^{b}	7.87 ± 0.32^{b}	5.88 ± 0.47^{a}
Aroma	8.19 ±0.24	8.07 ± 0.21	8.04 ± 0.12	8.03 ± 0.18	7.98 ± 0.16
Hardness	8.10 ±0.20	7.90 ± 0.21	7.87 ± 0.19	7.86 ± 0.10	7.77 ± 0.25
Mouthfeel	8.12 ±0.23^{b}	7.93 ± 0.33^{b}	7.90 ± 0.31^{b}	7.88 ± 0.16^{b}	6.91 ± 0.41^{a}
Overall acceptability	8.17 ±0.19^{b}	8.18 ± 0.16^{b}	8.17 ± 0.21^{b}	7.85 ± 0.27 ^{b}	6.22 ± 0.27^{a}

Recommendation

Composite meat chocolate can be successfully prepared by incorporating animal fat (10 per cent of total fat), dehydrated chicken meat powder (10 per cent), calcium lactate (1 per cent) and fruit extracts. Treatment of the product with the blueberry and raspberry extracts up to 3 per cent Level may improve the sensory attributes of calcium fortified composite meat chocolate.

Benefit/Impact

Incorporation of dehydrated ground chicken meat will provide a good source of protein and it will ultimately enhance the nutritional quality of chocolate. It helps in eviting the current problem of protein malnutrition in humans especially kids. Also the incorporation of calcium lactate will prevent the usual problem of dental caries in children who consume a lot of chocolates. Calcium lactate also have preservative action, thus it will improve the keeping quality of composite meat chocolate.

99. Fruit Fiber Fortified Emulsion based Spent Hen Meat Products (Patties)

Contact Scientist: **Dr. Sunil Kumar**

Introduction

Meat consist of almost all nutrients required essentially for human nutrition except the fiber content. Fiber is required for proper bowel movement and efficient assimilation of meat foods. Spent hen meat is undesirable due to poor texture quality.

Method

Fruits fiber after extraction of juice which is considered as waste is used as fiber source. The spent hen meat (≥72 weeks) was used for preparation of meat patties. The product prepared were analysed on physic-chemical, storage and sensory parameters.

Table 1: Formulation of Patties from Meat of Spent Hens

Parameters	*Fruit Fiber Fortified Spent Hen Meat Products (wt./wt.)*			
	Control	*5 per cent Orange Pulp*	*10 per cent Apple Pulp*	*10 per cent Pear Pulp*
Lean Meat	67.7	61.7	56.7	56.7
Water/crushed Ice	10.0	10.0	10.0	10.0
Vegetable oil	9.0	9.0	9.0	9.0
Condiment mixture	5.0	5.0	5.0	5.0
Refined Wheat Flour	4.0	4.0	4.0	4.0
Spice mixture	2.0	2.0	2.0	2.0
Table Salt	1.5	1.5	1.5	1.5
Carragennan	-	1.0	1.0	1.0
Fruit fiber	-	5.0	10.0	10.0
MSG	0.5	0.5	0.5	0.5
Sodium Tripolyphosphates	0.3	0.3	0.3	0.3
Sodium Nitrite	120ppm	120ppm	120ppm	120ppm

Flow Diagram for Preparation of Patties

Dressed spent hens

↓

Removal of skin and fat

↓

Deboning

↓

Cutting deboned meat into smaller chunks

↓

Mincing through 6mm plate

↓

Addition of salt (1.5 per cent), STPP (0.3 per cent) and nitrite (120 ppm)

↓

Chopping for 1.5 minute

↓

Addition of water (10 per cent) in the form of ice flakes

↓

Chopping for 1 minute

↓

Addition of vegetable oil (9 per cent)

↓

Chopping for 1-2 minute

↓

Addition of spice mixture, condiments mixture and refined wheat flour

↓

Chopping for 1.5-2 minutes

↓

Emulsion/patty mixture

↓

Moulding to disc shape in petriplates

↓

Cooking by oven roasted method

↓

Analysis of samples ⟵ Packaging ⟵ Cooling ⟶ Sensory evaluation

Table 2: Effect of Fruit Fiber on pH, Emulsion Stability and Proximate Composition of Oven Roasted Chicken Patties (Mean ±SE)*

Parameters	*Fruit Fiber Fortified Spent Hen Meat Products*			
	Control	*5 per cent Orange Pulp*	*10 per cent Apple Pulp*	*10 per cent Pear Pulp*
pH	6.34^{a} ±0.04	6.26^{ab} ±0.02	6.23^{b} ±0.01	6.13^{c}n±0.01
Cookong Yield (per cent)	91.33^{a} ±0.39	89.63^{b} ±0.51	88.07^{bc} ±0.40	87.56^{c} ±0.76
Moisture (per cent)	61.99^{a} ±0.32	60.67^{b} ±0.26	60.41^{bc} ±0.65	59.36^{c} ±0.36
Protein (per cent)	17.29^{a} ±0.23	16.47^{ab} ±0.36	16.42^{b} ±0.23	15.11^{c} ±0.26
Fat (per cent)	12.07^{a} ±0.25	11.23^{b} ±0.15	10.47^{c} ±0.23	10.29^{c} ±0.16
Crude Fibre(per cent)	0.71^{a} ±0.07	1.58^{b} ±0.16	2.48^{c} ±0.17	3.23^{d} ±0.08
Ash (per cent)	2.84^{a}±0.05	2.65^{b}±0.07	2.48^{bc}±0.03	2.33^{c} ±0.06
Moisture: Protein	3.58^{a}±0.04	3.68^{a}±0.08	3.69^{a}±0.05	3.93^{b}±0.06

*Mean ±SE with different superscripts in a row differs significantly ($P<0.05$). n = 6 for each treatment.

Table 3: Effect of Fruit Fiber on Sensory Quality Parameters of Oven Roasted Chicken Patties (Mean ±SE)*

Parameters	*Fruit Fiber Fortified Spent Hen Meat Products*			
	Control	*5 per cent Orange Pulp*	*10 per cent Apple Pulp*	*10 per cent Pear Pulp*
Appearance	6.70^{a} ±0.09	7.00^{b}±0.07	7.03^{b}±0.04	6.89^{a} ±0.09
Flavour	6.74±0.10	6.87±0.09	7.01±0.11	6.94±0.10
Juiciness	6.73^{a}±0.07	6.97^{b}±0.09	7.05^{b}±0.07	6.86^{a}±0.07
Texture	6.66^{a}±0.11	6.99^{ab}±0.10	7.06^{ab}±0.12	6.86^{a}±0.12
Overall acceptability	6.85^{a}±0.14	6.99^{a}±0.15	7.13^{a}±0.13	6.96^{a}±0.11

*Mean ±SE with different superscripts in a row differs significantly ($P<0.05$). Mean values are scores on 8 point descriptive scale where 1- extremely poor and 8- extremely desirable. = 21 for each treatment

Recommendation

The addition of apple pulp, orange pulp and pear pulp ($P<0.05$) decreased the emulsion stability and cooking yield but the addition of carrageenan significantly ($P<0.05$) increased the yield and emulsion stability. Pear pulp, apple pulp and orange pulp at 10 per cent, 10 per cent and 5 per cent levels respectively can be utilized for preparation of chicken patties from spent hen meat. Carrageenan at the level of 1 per cent could be used in fibre fortified patties. The prepared fruit fiber fortified Patties can be prepared without affecting its physic0chemical and sensory properties with shelf-life of more than fortnight.

Impact

1. The spent hen meat/Layers (≥72 weeks) can be used for preparation of meat product which otherwise is undesirable due to its coarse and tough texture.
2. The fruit fiber extract which is considered as waste, can be used as a fire source in meat products.
3. The prepared product was found to be having better shelf life than usual control product.

toxic effects of SLE supplementation has been observed in control rats as all the physiological (body weight, feed and water intake, organ to body weight ratio), biochemical, enzymatic parameters and oxidative stress markers did not showed any deviation from the normal control animals.

Further, in the study of effect of SLE in experimentally induced diabetes mellitus, SLE treatment of diabetic rats resulted in significant lowering of blood glucose levels and concomitant increase hepatic glycogen in diabetic rats in this study. Administration of SLE in diabetic rats also led to an improvement in other biochemical parameters such as TP, albumin, cholesterol, glucose, urea and creatinine compared to diabetic control group. Normalization of levels of various hepatic enzymes (AST, ALT, ACP and ALP) in serum and various oxidative stress markers in hepatic tissue by SLE were also observed. SLE administration to diabetic rats also resulted in improvement of various oxidative stress markers such as LPO, reduced glutathione, SOD and CAT. In the present study no toxic effects of oral administration of SLE has been observed in control rats as all the parameters including oxidative stress markers did not showed any deviation from the normal control animals.

Recommendations

1. Aqueous extract of Seabuckthorn leaves can be used effectively to ameliorate as well to protect from lead toxicity and its related complications.
2. Aqueous extract of Seabuckthorn leaves can be used to ameliorate diabetes related hyperglycemia and oxidative stress.

Benefits/Impact

1. Considering that lead toxicity is currently one of the serious problems worldwide, there is still no specific, reliable and safe treatment. Therefore there is need to look for an alternative solution to deal with lead poisoning such as "Antioxidant therapy". As many synthetic antioxidants have been shown to have one or the other side effects, there has been an upsurge of interest in the therapeutic potential of medicinal plants as antioxidants in reducing free radical induced tissue injury. Seabuckthorn leaves being the rich source of several antioxidants may provide an alternative solution to deal with lead poisoning.
2. Diabetes mellitus is ever increasing problem in India with 50.8 million diabetics. Many pharmaceutical drugs are presently available for the treatment of diabetes mellitus, however due to accompanying side effects, the use of these drugs are restricted. Seabuckthorn leaves extract plants may provide the useful source of new oral hypoglycemic compounds for the development of pharmaceutical entities or as dietary adjunct to existing therapies.

Index